Textbook of Oral Microbiology

Textbook of Oral Microbiology

KMK Masthan MDS
Professor and Head
Department of Oral Pathology and Microbiology
Sree Balaji Dental College and Hospital
Chennai, TN

N Aravindha Babu MDS
Associate Professor
Department of Oral Pathology and Microbiology
Sree Balaji Dental College and Hospital
Chennai, TN

CBS

CBS Publishers & Distributors Pvt Ltd

New Delhi • Bengaluru • Pune • Kochi • Chennai

Textbook of Oral Microbiology

ISBN: 978-81-239-1993-5

First Edition : 2011

Published by Satish Kumar Jain and produced by Vinod K. Jain for
CBS Publishers & Distributors Pvt Ltd
4819/XI Prahlad Street, 24 Ansari Road, Daryaganj
New Delhi 110 002, India.
Ph: 23289259, 23266861, 23266867
Fax: 011-23243014

Website: www.cbspd.com
e-mail: delhi@cbspd.com
cbspubs@vsnl.com
cbspubs@airtelmail.in.

Branches

- Bengaluru: Seema House 2975, 17th Cross, K.R. Road, Banasankari 2nd Stage, Bengaluru 560 070, Karnataka
  Ph: +91-80-26771678/79 Fax: +91-80-26771680 e-mail: bangalore@cbspd.com
- Pune: Bhuruk Prestige, Sr. No. 52/12/2+1+3/2 Narhe, Haveli (Near Katraj-Dehu Road Bypass), Pune 411 051, Maharashtra
  Ph: 020-64704058, 64704059, 32342277 Fax: +91-020-24300160 e-mail: pune@cbspd.com
- Kochi: 36/14 Kalluvilakam, Lissie Hospital Road, Kochi 682 018, Kerala
  Ph: +91-484-4059061-65 Fax: +91-484-4059065 e-mail: cochin@cbspd.com
- Chennai: 20, West Park Road, Shenoy Nagar, Chennai 600 030, Tamil Nadu
  Ph: +91-44-26260666, 26208620 Fax: +91-44-45530020 email: chennai@cbspd.com

Printed at Manipal Press Limited, Manipal, Karnataka

to

---

the management of Sree Balaji Dental College and Hospital who have helped me to pursue academics and writing for the past 17 years

Foreword

VK Subburaj IAS
Principal Secretary to Government

Health and Family Welfare Department
Secretariat, Chennai 600 009
Off ☎ : 25671875
Fax : 2567 1253
E-mail : hfsec@tn.gov.in

It is with great pleasure that I write this Foreword since I sense that the text of this book unfolds the mystery of how microbes produce common dental diseases. Dentistry has always fascinated me with its peculiar diseases and I have come to realize through my involvement with numerous public health camps and projects that majority of the population suffer from these dental diseases.

From my interaction with several health professionals in the dental education side and the field work side, I was able to understand that the phenomenal frequency with which the dental diseases affect the population will be a mammoth challenge for the available resources in the form of available manpower and existing technology to handle. Hence dental education has to be focused and the curriculum geared up to expand the knowledge of the students of dentistry.

Books of this kind which clearly focus on the common dental diseases are the need of the hour. A good teacher can educate probably fifty or sixty students every year whereas a good book, though it may not have the impact of a teacher, can help thousands of students over a period of several years. Since both the authors, Dr KMK Masthan MDS and Dr N Aravindha Babu MDS, who have written this book are well versed academicians with many books to their credit both in English and Tamil, I hope they have poured in their best possible effort, experience and knowledge into this book.

VK Subburaj IAS
Principal Secretary to Government
Health and Family Welfare Department
Government of Tamil Nadu

Preface

The two commonest oral diseases, namely periodontitis and dental caries, are microbial diseases. Most lesions in the oral cavity arise because of infections by bacteria, viruses and fungi. In clinical dental practice ninety percent of the time of general dentist is spent to treat these lesions. Still oral microbiology is one field which does not get its due share in the dental student's study portfolio. The main reason is this vast and very important subject is taught as a part of oral pathology where neoplasia, cysts and developmental disturbances dominate the attention of the student. As a result, the diseases which are rare and most of which can only be identified and for which no treatment can be done by the general dentist occupy his study hours and examinations are also more focused on these diseases only. What use is it if you cannot treat headache but you can identify a craniofacial anomaly.

Since we felt strongly about this aspect of dental curriculum, we have tried to compile relevant literature on oral microbes and the diseases caused by them. This is not a journal where original research is documented and claimed. Here the target reader is the dental student in his/her second to fifth year and, of course, any other medical or dental professional who is interested.

A fairly comprehensive but not an exhaustive attempt has been made to cover all aspects of common oral microbial diseases. A lot was written but some were deleted later since we did not want to frighten a BDS student with a voluminous book. Hope our decisions regarding which were important for the young student are acceptable to the veterans in the field of microbiology and oral pathology. Criticism and constructive comments are welcome at masthankmk@yahoo.com.

I am greatly indebted to Dr S Satheesh and Dr L Malathy, both postgraduate students of oral pathology and microbiology at Sree Balaji Dental College, for their help in compiling the data.

My wholehearted thanks to CBS P&D, doyens in the field of medical publications, for taking up our work for publication.

KMK Masthan
N Aravindha Babu

Contents

1 Normal Oral Flora and Oral Pathogens

ORAL FLORA

Bacteria that reside in the oral cavity occupy a unique and fascinating habitat. The mouth is the only part of the body where hard tissues are naturally exposed to the external environment. Bacteria firmly adhere to and grow on the tooth to form a complex biofilm known as dental plaque. The complexity of the oral microbial ecology is magnified by the fact that the mouth also possesses a variety of other surfaces, including the buccal and vestibular mucosa, hard palate, tongue, and the floor of the mouth, all which provide unique habitats for microbial colonization.

The indigenous microbiota plays an important role in health and diseases of humans and animals. It contributes to the development of the immune system and provides resistance to colonization by allochthonous or pathogenic microorganisms. It also constitutes a reservoir of potentially pathogenic bacteria that may infect host tissues.

In the oral cavity, indigenous bacteria are often associated with the etiology of two major oral diseases, which are endemic in industrialized societies and are increasing in developing countries. Oral diseases seem to appear after an imbalance among the indigenous microbiota, leading to the emergence of potentially pathogenic bacteria. To define the process involved in caries and periodontal diseases, it is necessary to understand the ecology of the oral cavity and to identify the factors responsible for the transition of the oral microbiota from a commensal to a pathogenic relationship with the host. The regulatory forces influencing the oral ecosystem can be divided into three major categories: Host related, microbe related, and external factors.

Secretory immunoglobulin A (SIgA) constitutes the predominant immunoglobulin isotype in secretions, including saliva. It is considered to be the first line of defense of the host against pathogens which colonize or invade surfaces bathed by external secretions. The main function of SIgA antibodies seems to be limit microbial adherence as well as penetration of foreign antigens into the mucosa. Naturally occurring SIgA antibodies reactive with a variety of indigenous bacteria have been detected in saliva. Furthermore, indigenous bacteria of the oral cavity have been found to be coated with SIgA. The role of these antibodies in the colonization and the regulation of the indigenous microbiota is still controversial.

Despite the presence of SIgA antibodies, a resident microbiota persists in the oral cavity. Indigenous bacteria can survive in the oral cavity because they are less susceptible to or can avoid immune mechanisms. It is also possible that SIgA has an effect on indigenous bacteria but that it is only a minor force among the multiple factors that maintain the homeostasis of the indigenous microbiota.

Since caries and periodontal diseases are associated with indigenous bacteria, defining the role of SIgA in the control of the oral indigenous microbiota is a prerequisite for the elaboration of effective vaccines against these diseases. In vitro experiments have shown that SIgA may inhibit or promote the adherence of oral bacteria to teeth. Experiments with animal models showed that salivary IgA induced against *Streptococcus mutans* leads to a reduction in the colonization of this bacterium and to the prevention of caries. More recent studies indicate that the immunity is not maintained. IgA-deficient humans were found to be more or less susceptible to caries and periodontal diseases.

The present review describes the oral ecosystems, the major factors that might control the oral microbiota, the basic aspects of the secretory immune system, and the biological functions of SIgA and finally focuses on experiments related to the role of IgA in the microbial ecology of the oral cavity. To present an overall picture of the current knowledge of oral microbial ecology, this review is not limited to human studies but includes in vitro systems such as the chemostat and the artificial mouth.

Ecology is the science that studies interrelationships between organisms and their living (biotic) and nonliving (abiotic) environment.

An ecosystem consists of the microbial community living in a defined habitat and the abiotic surroundings composed of physical and chemical elements. In its simplest expression, the oral ecosystem is thus composed of the oral microorganisms and their surroundings, the oral cavity.

Within an ecosystem, the development of a community usually involves a succession of population. The process begins with the colonization of the habitat by pioneer microbial population. In the oral cavity of newborns, streptococci (*S. mitis* biovar 1, *S. oralis*, and *S. salivarius*) are the pioneer organisms. Pioneer microorganisms fill the niche of this new environment and modify the habitat, and as a result, new population may develop. As the process continues, the diversity and the complexity of the microbial community increased. Succession ends when no additional niche is available for new population. At this stage, a relatively stable assemblage of bacterial populations is achieved. It is called a climax community.

The concept of a stable or climax community does not imply static conditions. The stability is based upon homeostasis, which implies compensating mechanisms that act to maintain steady-state conditions by a variety of controls aimed at counteracting perturbations that would upset the steady state. The concepts of homeostasis and bacterial succession are important in oral microbiology. Some factors, such as a high-sucrose diet, may cause an irreversible breakdown in the homeostasis of the oral ecosystem, resulting in the initiation of caries.

NORMAL FLORA

The predominant bacterial floras of humans are shown in Table 1.1. This table lists only a fraction of the total bacterial species that occur as normal flora of humans.

In a healthy individuals, the internal tissues, e.g. blood, brain, muscle, are normally free of microorganisms. However, the surface tissues, i.e. skin and mucous membranes are constantly in contact with environmental organisms and become readily colonized by

Table 1.1: Bacteria commonly found on the surfaces of the human body

Bacterium	*Skin*	*Conjunctiva*	*Nose*	*Pharynx*	*Mouth*	*Lower GI*	*Ant. urethra*	*Vagina*
Staphylococcus epidermidis (1)	++	+	++	++	++	+	++	++
*Staphylococcus aureus** (2)	+	+/-	+	+	+	++	+/-	+
Streptococcus mitis				+	++	+/-	+	+
Streptococcus salivarius				++	++			
*Streptococcus mutans** (3)				+	++			
*Enterococcus faecalis** (4)				+/-	+	++	+	+
*Streptococcus pneumoniae** (5)		+/-	+/-	+	+			+/-
*Streptococcus pyogenes** (6)	+/-	+/-		+	+	+/-		+/-
Neisseria sp. (7)		+	+	++	+		+	+
*Neisseria meningitidis** (8)			+	++	+			+
*Enterobacteriaceae**	+/-	+/-	+/-	+	++	+	+	
(*Escherichia coli*) (9)								
Proteus sp.		+/-	+	+	+	+	+	+
*Pseudomonas aeruginosa** (10)				+/-	+/-	+	+/-	
*Haemophilus influenzae** (11)		+/-	+	+	+			
Bacteroides sp.*						++	+	+/-
Bifidobacterium bifidum (12)						++		
Lactobacillus sp. (13)				+	++	++		++
Clostridium sp.* (14)					+/-	++		
Clostridium tetani (15)						+/-		
Corynebacterium (16)	++	+	++	+	+	+	+	+
Mycobacterium	+		+/-	+/-		+	+	
Actinomycetes				+	+			
Spirochetes				+	++	++		
Mycoplasmas				+	+	+	+/-	+

++ = nearly 100 percent + = common (about 25 percent) +/- = rare (less than 5%) * = potential pathogen

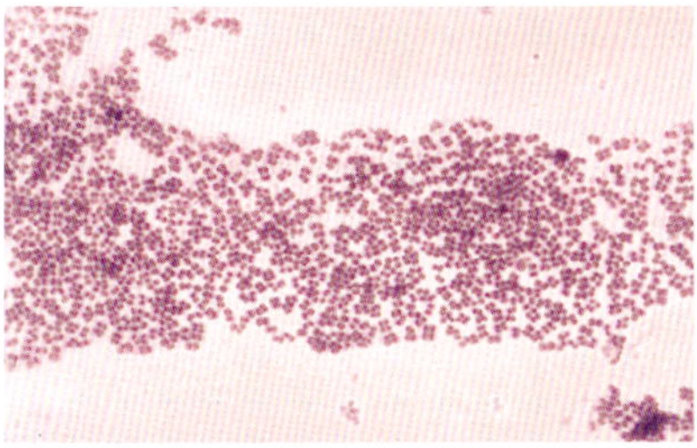

Gram stain of a species of *Micrococcus*, commonly isolated from the skin and nasal membranes of humans

various microbial species. The mixture of organisms regularly found at any anatomical site is referred to as the normal flora.

16S RNA probes to survey the diversity of bacteria in dental plaque revealed that only one percent of the total species found have ever been cultivated. Similar observations have been made with the intestinal flora.

Table 1.1 Notes

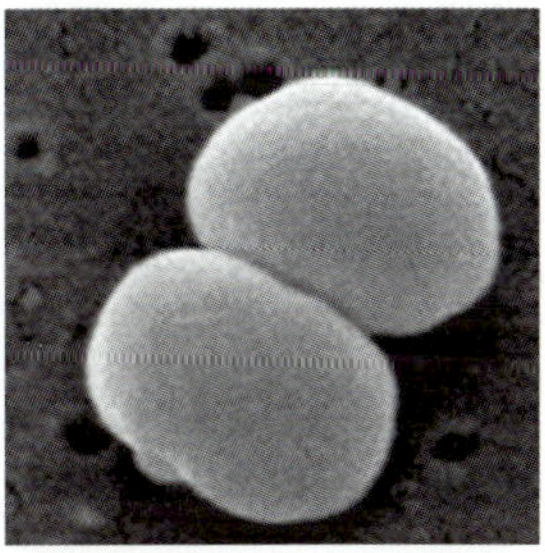

Staphylococcus epidermidis. Scanning EM

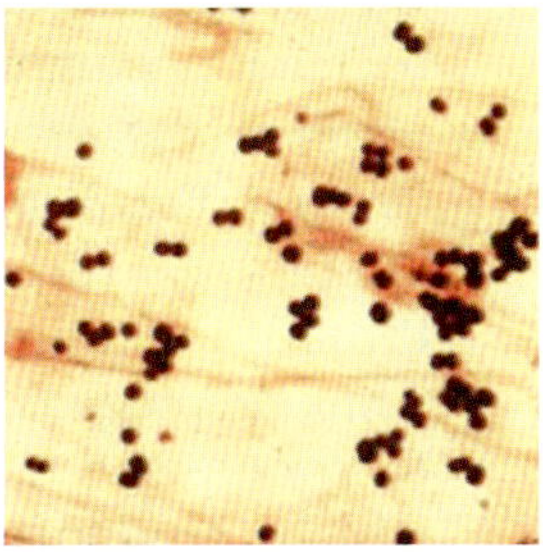

S. aureus. Gram stain

1. *Staphylococcus epidermidis* is highly adapted to the diverse environments of its human host. *S. aureus* is a potential pathogen. It is a leading cause of bacterial disease in humans. It can be transmitted from the nasal membranes of an asymptomatic carrier to a susceptible host.
2. Many of the normal flora are either pathogens or opportunistic pathogens. The asterisks indicate members of the normal flora that may be considered major pathogens of humans.
3. *Streptococcus mutans* is the primary bacterium involved in plaque formation and initiation of dental caries. Viewed as an opportunistic infection, dental disease is one of the most prevalent and costly infectious diseases in the United States.

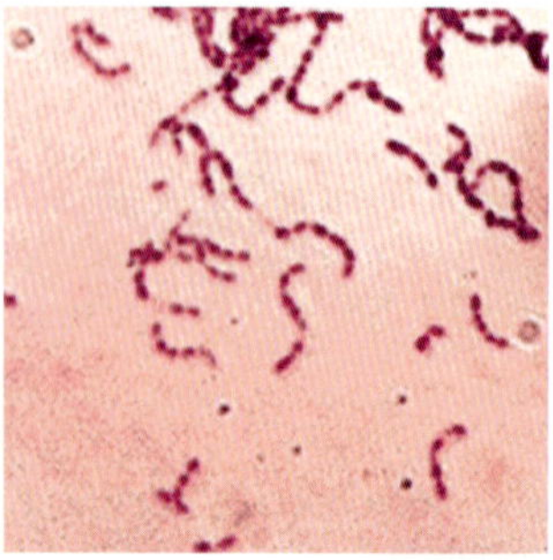

Streptococcus mutans. Gram stain

4. *Enterococcus faecalis* was formerly classified as *Streptococcus faecalis*. The bacterium is such regular component of the intestinal flora, that many European countries use it as the standard indicator of fecal pollution, *Enterococcus faecalis* has emerged as a significant, antibiotic-resistant, nosocomial pathogen.

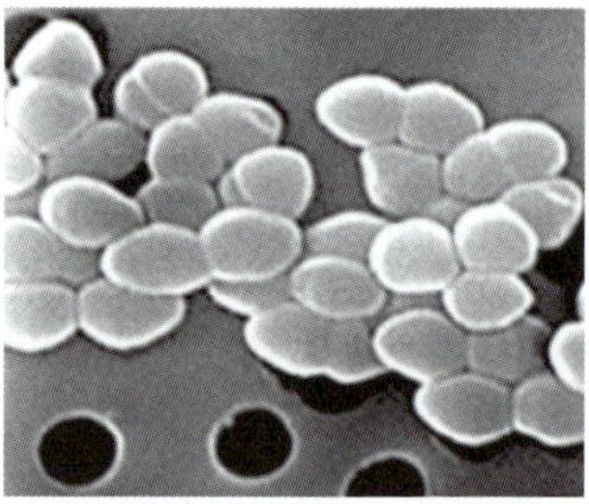

Vancomycin Resistant *Enterococcus faecalis*. Scanning EM

5. *Streptococcus pneumoniae* is present in the upper respiratory tract of about half the population. If it invades the lower respiratory tract it can cause pneumonia. *Streptococcus pneumoniae* causes 95 percent of all bacterial pneumonia.
6. *Streptococcus pyogenes* refers to the Group A, Beta-hemolytic streptococci. Streptococci cause tonsillitis (strep throat), pneumonia, endocarditis. Some streptococcal diseases can lead to rheumatic fever or nephritis which can damage the heart and kidney.

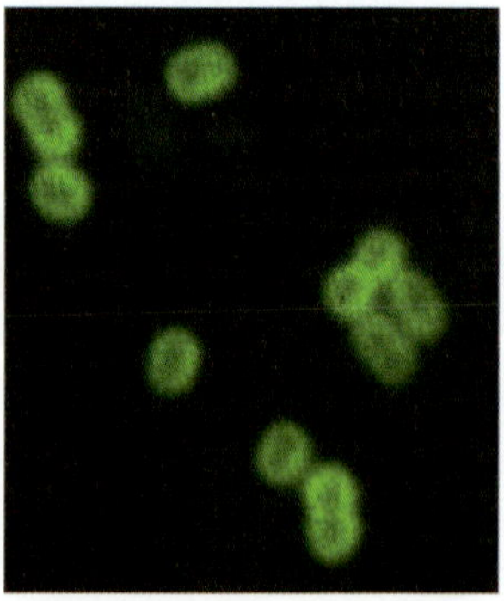

Streptococcus pneumoniae. Direct fluorescent antibody stain

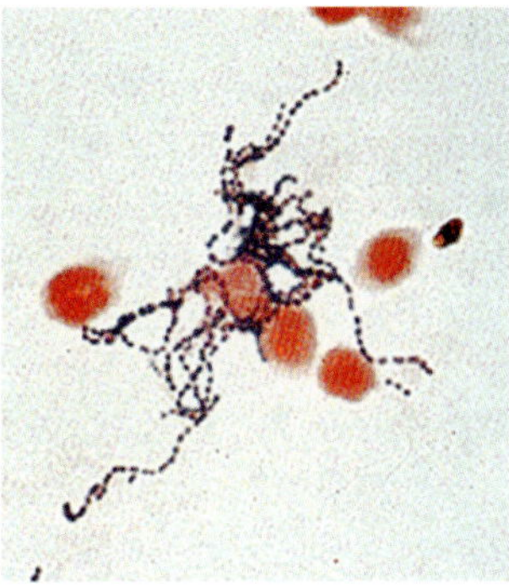

Streptococcus pyogenes. Gram stain

7. *Neisseria* and other gram-negative cocci are frequent inhabitants of the upper respiratory tract, mainly the pharynx. *Neisseria meningitidis,* an important cause of bacterial meningitis that can colonize as well until the host can develop active immunity against the pathogen.
8. While *E. coli* is a consistent resident of the small intestine, many other enteric bacteria may reside here as well, including *Klebsiella, Enterobacter* and *Citrobacter*. Some strains of *E. coli* are pathogens that cause intestinal infections, urinary tract infections and neonatal meningitis.

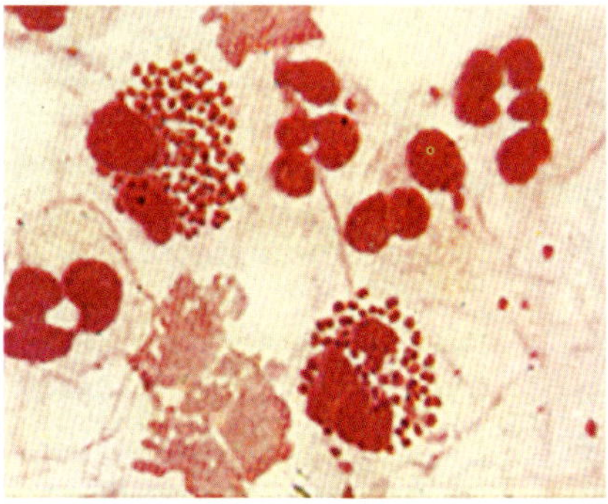

Neisseria meningitidis. Gram stain

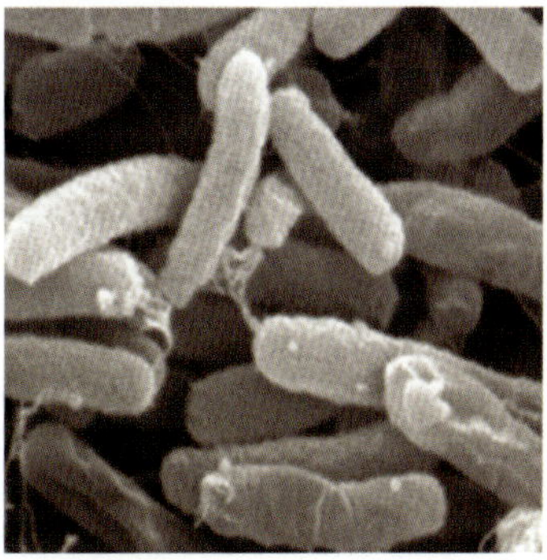

E. coli. Scanning EM

9. *Pseudomonas aeruginosa* is the quintessential opportunistic pathogen of humans that can invade virtually any tissue. It is a leading cause of hospital-acquired (nosocomial) Gram-negative infections, but its source is often exogenous (from outside the host).

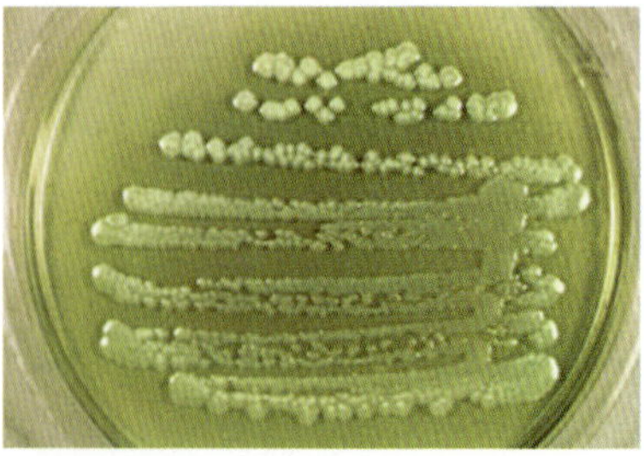

Colonies of *Pseudomonas aeruginosa* growing on an agar plate. The most virulent *Pseudomonas* species produce mucoid colonies and green pigments such as this isolate

10. *Haemophilus influenzae* is a frequent secondary invader to viral influenza, and was named accordingly. The bacterium was the leading cause of meningitis in infants and children until the recent development of the *H. influenzae* type B vaccine.

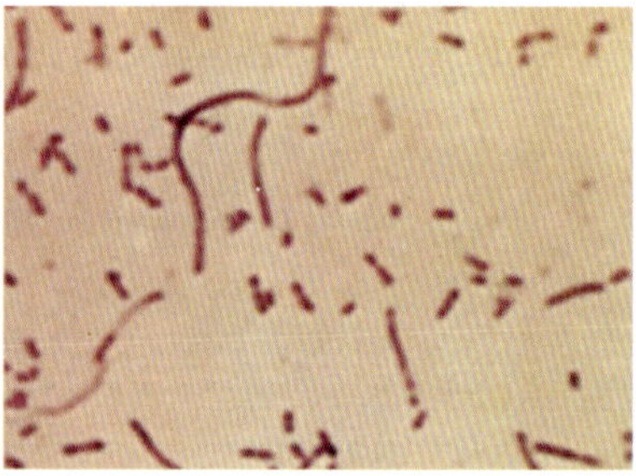

Haemophilus influenzae. Gram stain

11. The greatest numbers of bacteria are found in the lower intestinal tract, specifically the colon and the most prevalent bacteria are the *Bacteroides*, a group of gram-negative, anaerobic, non-spore forming bacteria. They have been implicated in the initiation of colitis and colon cancer.

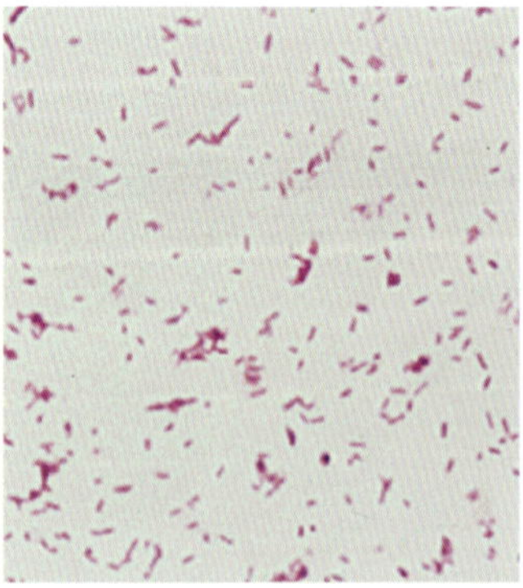

Bacteroides fragilis. Gram stain

12. Bifidobacteria are Gram-positive, non-sporeforming, lactic acid bacteria. They have been described as "friendly" bacteria in the intestine of humans. *Bifidobacterium bifidum* is the predominant bacterial species in the intestine of breast-fed infants, where it presumably prevents colonization by potential pathogens. These bacteria are sometimes used in the manufacture of yogurts and are frequently incorporated into probiotics.

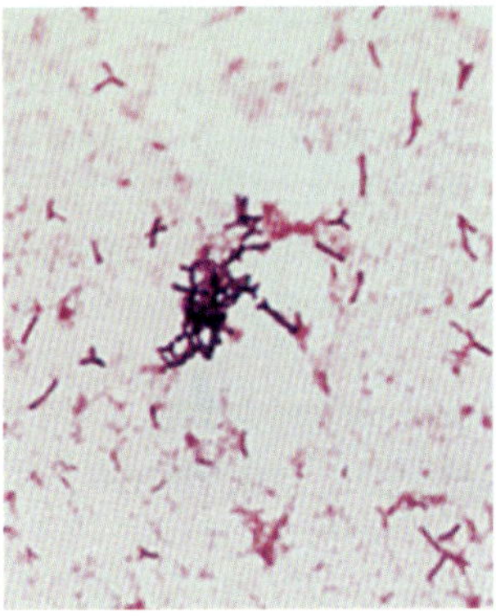

B. bifidum. Gram stain.

13. Lactobacilli in the oral cavity probably contribute to acid formation that leads to dental caries. *Lactobacillus acidophilus* colonizes the vaginal epithelium during child-bearing years and establishes the low pH that inhibits the growth of pathogens.
14. There are numerous species of *Clostridium* that colonize the bowel. *Clostridium perfringens* is commonly isolated from feces. *Clostridium difficile* may colonize the bowel and cause "antibiotic-induced diarrhea" or pseudomembranous colitis.
15. *Clostridium tetani* is included in the table as an example of a bacterium that is "transiently associated" with humans as a

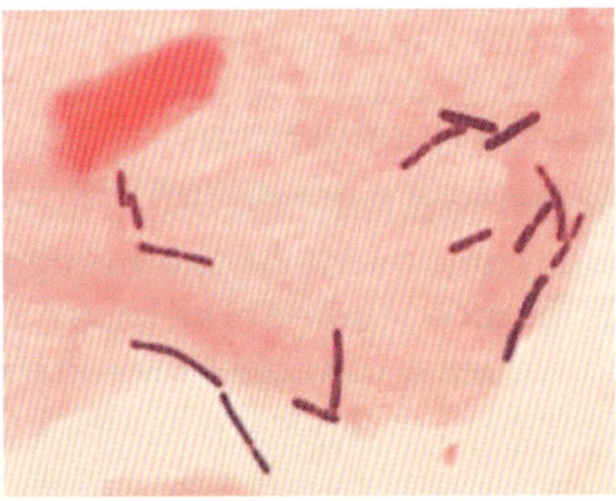

Lactobacillus species and a vaginal squamous epithelial cell

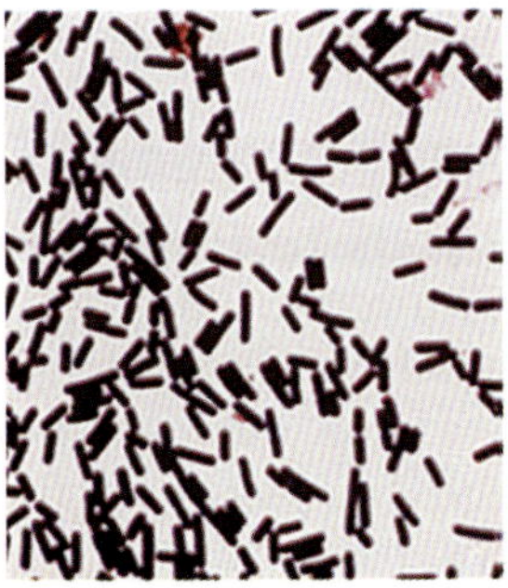

C. perfringens. Gram stain

component of the normal flora. The bacterium can be isolated from feces in 0–25 percent of the population. The endospores are probably ingested with food and water, and the bacterium does not colonize the intestine.

16. The corynebacteria, and certain related propionic acid bacteria, are consistent skin flora. Some have been implicated as a cause of acne. *Corynebacterium diphtheriae, the* agent of diphtheria, was considered a member of the normal flora before the widespread use of the diphtheria toxoid, which is used to immunize.

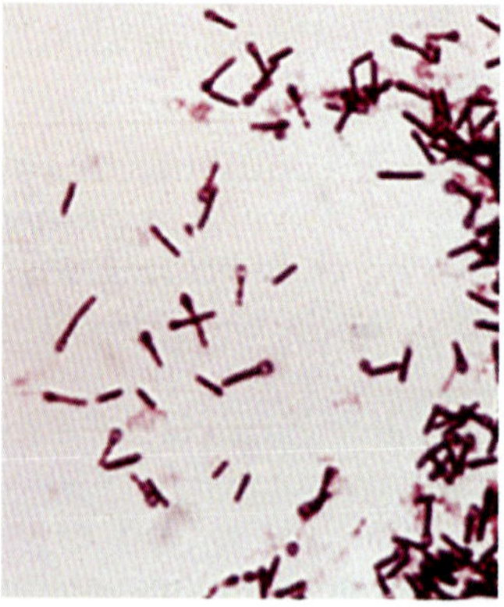

Clostridium tetani. Gram stain

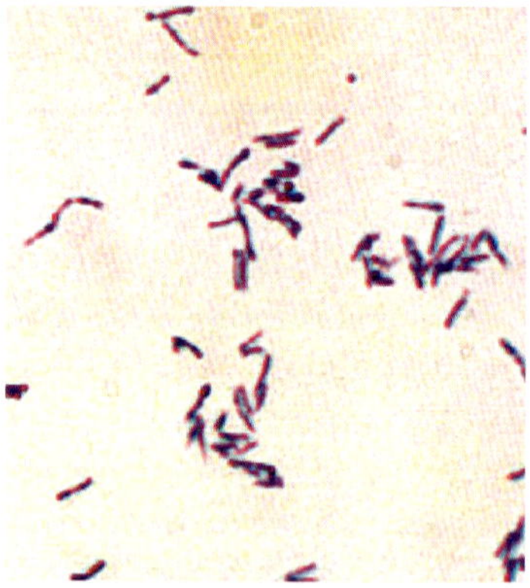

Against the disease. *Corynebacterium diphtheriae.* No longer a part of the normal flora

Associations between Human and the Normal Flora

E. coli is the best known bacterium that regularly associates itself with humans, being an invariable component of the human intestinal tract. Even though *E. coli* is the most studied of all bacteria, and we know the exact location and sequence of 4, 288 genes on its chromosome, we do not fully understand its ecological relationship with humans.

In fact, not much is known about the nature of the associations between humans and their normal flora, but they are thought to be dynamic interactions rather than associations of mutual indifference. Both host and bacteria are thought to derive benefit from each other, and the associations are, for the most part, **mutualistic**. The normal flora derives from their host a steady supply of nutrients, a stable environment, and protection and transport. The host obtains from the normal flora certain nutritional and digestive benefits, stimulation of the development and activity of immune system, and protection against colonization and infection by pathogenic microbes.

While most of the activities of the normal flora benefit their host, some of the normal flora are **parasitic** (live at the expense of their host), and some are **pathogenic** (capable of producing disease). Diseases that are produced by the normal flora in their host may be called **endogenous diseases**. Most endogenous bacterial diseases are **opportunistic infections**, meaning that the organism must be given a special opportunity of weakness or let-down in the host defences in order to infect. An example of an opportunistic infection is chronic bronchitis in smokers wherein normal flora bacteria are able to invade the weakened lung.

Sometimes the relationship between a member of the normal flora and its host cannot be deciphered. Such a relationship where there is no apparent benefit or harm to either organism during their association is referred to as a **commensal relationship**. Many of the normal flora that are not predominant in their habitat, even though always present in low numbers, are thought of as commensal bacteria. However, if a presumed commensal relationship is studied in detail, parasitic or mutualistic characteristics often emerge.

Tissue Specificity

Most members of the normal bacterial flora prefer to colonize certain tissues and not others. This "tissue specificity" is usually due to properties of both the host and the bacterium. Usually, specific bacteria colonize specific tissues by one or another of these mechanisms.

1. **Tissue tropism** is the bacterial preference or predilection for certain tissues for growth. One explanation for tissue tropism is that the host provides essential nutrients and growth factors for

the bacterium, in addition to suitable oxygen, pH, and temperature for growth.

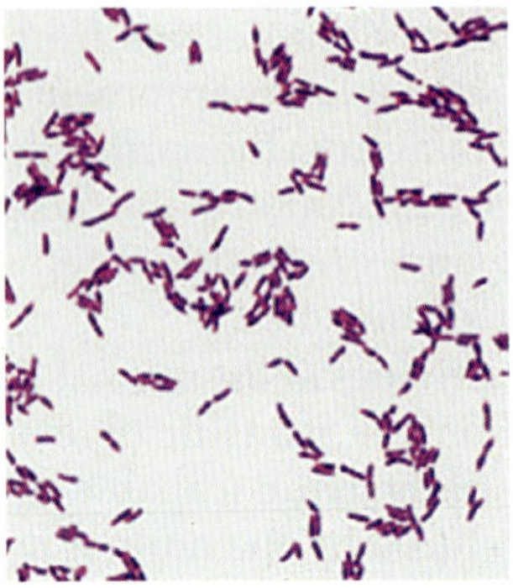

Lactobacillus acidophilus, informally known as "Doderlein's bacillus" colonizes the vagina because glycogen is produced which provides the bacteria with a source of sugar that they ferment to lactic acid

2. **Specific adherence:** Most bacteria can colonize a specific tissue or site because they can adhere to that tissue or site in a specific manner that involves complementary chemical interactions between the two surfaces. Specific adherence involves biochemical interactions between bacterial surface components (**ligands** or **adhesins**) and host cell molecular **receptors**. The bacterial components that provide adhesins are molecular parts

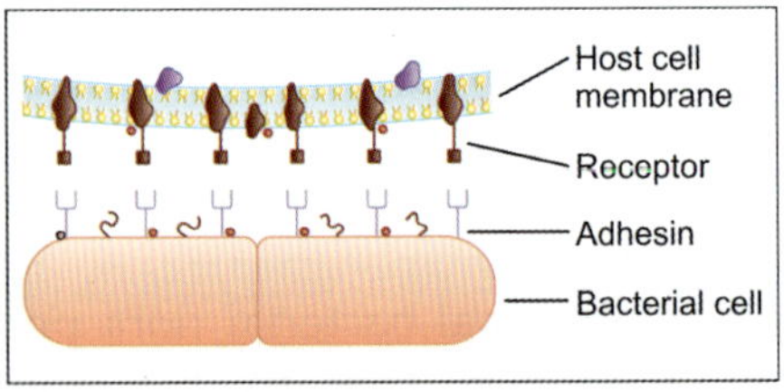

Specific adherence involves complementary chemical interactions between the host cell or tissue surface and the bacterial surface. In the language of medical microbiologist, a bacterial "adhesin" attaches covalently to a host "receptor" so that the bacterium "docks" itself on the host surface. The adhesins of bacterial cells are chemical components of capsules, cell walls, pili or fimbriae. The host receptors are usually glycoproteins located on the cell membrane or tissue surface

of their capsules, fimbriae, or cell walls. The receptors on human cells or tissues are usually glycoprotein molecules located on the host cell or tissue surface.

Some examples of adhesins and attachment sites used for specific adherence to human tissues are described in the Table 1.2.

Table 1.2: Examples of bacterial specific adherence to host cells or tissue

Bacterium	*Bacterial adhesin*	*Attachment site*
Streptococcus pyogenes	Cell-bound protein (M-protein)	Pharyngeal epithelium
Streptococcus mutans	Cell- bound protein (Glycosyl transferase)	Pellicle of tooth
Streptococcus salivarius	Lipoteichoic acid	Buccal epithelium of tongue
Streptococcus pneumoniae	Cell-bound protein (choline-binding protein)	Mucosal epithelium
Staphylococcus aureus	Cell-bound protein	Mucosal epithelium
Neisseria gonorrhoeae	N-methylphenyl-alanine pili	Urethral/cervical epithelium
Enterotoxigenic E. coli	Type-1 fimbriae	Intestinal epithelium
Uropathogenic E. coli	P-pili (pap)	Upper urinary tract
Bordetella pertussis	Fimbriae ("filamentous hemagglutinin")	Respiratory epithelium
Vibrio cholerae	N-methylphenylalanine pili	Intestinal epithelium
Treponema pallidum	Peptide in outer membrane	Mucosal epithelium
Mycoplasma	Membrane protein	Respiratory epithelium
Chlamydia	Unknown	Conjunctival or urethral epithelium

3. **Biofilm formation:** Some of the indigenous bacteria are able to construct **biofilms** on a tissue surface, or they are able to colonize a biofilm built by another bacterial species. Many biofilms are a

mixture of microbes, although one member is responsible for maintaining the biofilm and may predominate.

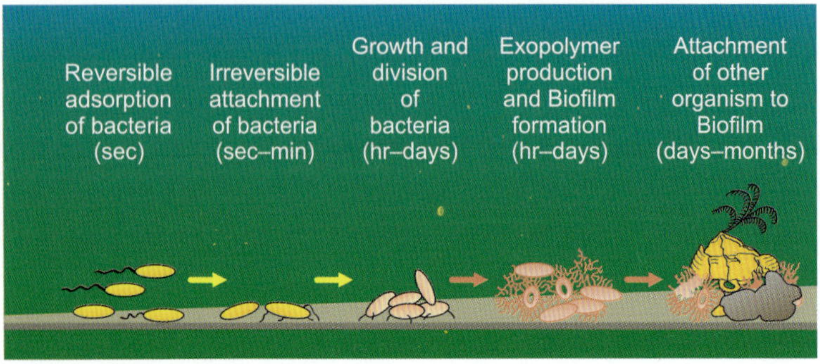

Cartoon depicting biofilm formation. Biofilms usually occur when one bacterial species attaches specifically or non specifically to a surface, and then secretes carbohydrate slime (exopolymer) that imbeds the bacteria and attracts other microbes to the biofilm for protection or nutritional advantages

The classic biofilm that involves components of the normal flora of the oral cavity is the formation of dental plaque on the teeth. Plaque is a naturally-constructed biofilm, in which the consortia of bacteria may reach a thickness of 300–500 cells on the surfaces of the teeth. These accumulations subject the teeth and gingival tissues to high concentrations of bacterial metabolites, which result in dental disease.

Composition of the Normal Flora

The normal flora of humans are exceedingly complex and consist of more than 200 species of bacteria. The makeup of the normal flora may be influenced by various factors, including genetics, age, sex, stress, nutrition and diet of the individual.

Three developmental changes in humans, weaning, the eruption of the teeth, and the onset and cessation of ovarian functions, invariably affect the composition of the normal flora in the intestinal tract, the oral cavity, and the vagina, respectively. However, within the limits of these fluctuations, the bacterial flora of humans is sufficiently constant to a give general description of the situation.

A human first becomes colonized by a normal flora at the moment of birth and passage through the birth canal. In utero, the fetus is sterile, but when the mother's water breaks and the birth process begins, so does colonization of the body surfaces. Handling and feeding of the infant after birth leads to establishment of a stable normal flora on the skin, oral cavity and intestinal tract in about 48 hours.

It has been calculated that a human adult houses about 10^{12} bacteria on the skin, 10^{10} in the mouth, and 10^{14} in the gastrointestinal tract. The latter number is far in excess of the number of eucaryotic cells in all the tissues and organs which comprise a human. The predominant bacteria on the surfaces of the human body are listed in Table 1.3.

Table 1.3: Predominant bacteria at various anatomical locations in adults

Anatomical location	*Predominant bacteria*
Skin	Staphylococci and corynebacteria
Conjunctiva	Sparse Gram-positive cocci and Gram-negative rods
Oral cavity	
Teeth	Streptococci, lactobacilli
Mucous membranes	Streptococci and lactic acid bacteria
Upper respiratory tract	
Nares (nasal membranes)	Staphylococci and corynebacteria
Pharynx (throat)	Streptococci, neisseria, Gram-negative rods and cocci
Lower respiratory tract	None
Gastrointestinal tract	
Stomach	Helicobacter pylori (up to 50%)
Small intestine	Lactics, enterics, enterococci, bifidobacteria
Colon	Bacteroides, lactics, enterics, enterococci, clostridia, methanogens
Urogenital tract	
Anterior urethra	Sparse staphylococci, corynebacteria enterics
Vagina	Lactic acid bacteria during child-bearing years; otherwise mixed

Normal Flora of the Skin

The adult human is covered with approximately 2 square meters of skin. The density and composition of the normal flora of the skin varies with anatomical locale. The high moisture content of the axilla, groin, and areas between the toes supports the activity and growth of relatively high densities of bacterial cells, but the density of bacterial populations at most other sites is fairly low, generally in 100s or 1000s per square cm. Most bacteria on the skin are sequestered in sweat glands.

The skin microbes found in the most superficial layers of the epidermis and the upper parts of the hair follicles are Gram-positive cocci (*Staphylococcus epidermidis* and *Micrococcus* sp.) and corynebacteria such as *Propionibacterium* sp. These are generally nonpathogenic and considered to be commensal, although mutualistic and parasitic roles have been assigned to them. For example, staphylococci and propionibacteria produce fatty acids that inhibit the growth of fungi and yeast on the skin. But, if *Propionibacterium acnes,* a normal inhabitant of the skin, becomes trapped in hair follicle, it may grow rapidly and cause inflammation and acne.

Sometimes potentially pathogenic *Staphylococcus aureus* is found on the face and hands in individuals who are nasal carriers. This is because the face and hands are likely to become inoculated with the bacteria on the nasal membranes. Such individuals may autoinoculate themselves with the pathogen or spread it to other individuals or foods.

Normal Flora of the Conjunctiva

A variety of bacteria may be cultivated from the normal conjunctiva, but the number of organisms is usually small. *Staphylococcus epidermidis* and certain coryneforms (*Propionibacterium acnes*) are dominant. *Staphylococcus aureus, some* streptococci, *Haemophilus* sp. and *Neisseria* sp. are occasionally found. The conjunctiva is kept moist and healthy by the continuous secretions from the lacrimal glands. Blinking wipes the conjunctiva every few seconds mechanically washing away foreign objects including bacteria. Lacrimal secretions (tears) also contain bactericidal substances including lysozyme. There is little or no opportunity for microorganisms to colonize the conjunctiva without special mechanisms to attach to the epithelial surfaces and some ability to withstand attack by lysozyme.

Pathogens which do infect the conjunctiva (e.g. *Neisseria gonorrhoeae* and *Chlamydia trachomatis*) are thought to be able to specifically attach to the conjunctival epithelium. Newborn infants may be especially prone to bacterial attachment. Since *Chlamydia* and *Neisseria* might be present on the cervical and vaginal epithelium of an infected mother, silver nitrate or an antibiotic may be put into the newborn's eyes to avoid infection after passage through the birth canal.

Colonies of *Propionibacterium acnes*, found on skin and the conjunctiva

Normal Flora of the Respiratory Tract

A large number of bacterial species colonize the upper respiratory tract (nasopharynx). The nares (nostrils) are always heavily colonized, predominantly with *Staphylococcus epidermidis* and corynebacteria, and often (in about 20% of the general population) with *Staphylococcus aureus,* this being the main carrier site of this important pathogen. The healthy sinuses, in contrast are sterile. The pharynx (throat) is normally colonized by streptococci and various Gram-negative cocci. Sometimes pathogens such as *Streptococcus pneumoniae, Streptococcus pyogenes, Haemophilus influenzae* and *Neisseria meningitidis* colonize the pharynx.

The lower respiratory tract (trachea, bronchi, and pulmonary tissues) is virtually free of microorganisms, mainly because of the efficient cleansing action of the ciliated epithelium which lines the tract. Any bacteria reaching the lower respiratory tract are swept upward by the action of the mucociliary blanket that lines the bronchi, to be removed subsequently by coughing, sneezing, swallowing, etc. If the respiratory tract epithelium becomes damaged, as in bronchitis or viral pneumonia, the individual may become susceptible to infection by pathogens such as *H. influenzae* or *S. pneumoniae* descending from the nasopharynx.

Normal Flora of the Urogenital Tract

Urine is normally sterile, and since the urinary tract is flushed with urine every few hours, microorganisms have problems gaining access and becoming established. The flora of the anterior urethra, as indicated principally by urine cultures, suggests that the area may be inhabited by a relatively consistent normal flora consisting of *Staphylococcus epidermidis, Enterococcus faecalis* and some alpha-hemolytic streptococci. Their numbers are not plentiful, however. In addition, some enteric bacteria (e.g. *E. coli, Proteus*) and corynebacteria, which are probably contaminants from the skin, vulva or rectum, may occasionally be found at the anterior urethra.

The vagina becomes colonized soon after birth with corynebacteria, staphylococci, streptococci, *E. coli,* and a lactic acid bacterium historically named "Doderlein's bacillus" (*Lactobacillus acidophilus*). During reproductive life, from puberty to menopause, the vaginal epithelium contains glycogen due to the actions of circulating estrogens. Doderlein's bacillus predominates, being able

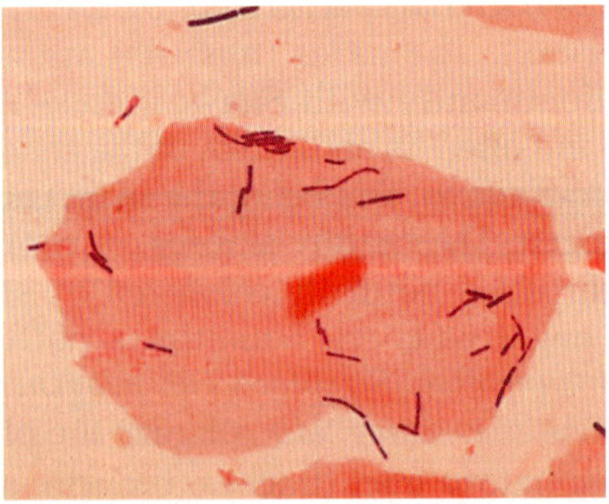

Lactobacillus acidophilus species, possibly Doderlein's bacillus, in association with a vaginal epithelial cell

to metabolize the glycogen to lactic acid. The lactic acid and other products of metabolism inhibit colonization by all except this lactobacillus and a select number of lactic acid bacteria. The resulting low pH of the vaginal epithelium prevents establishment by most other bacteria as well as the potentially-pathogenic yeast, *Candida albicans*. This is a striking example of the protective effect of the normal bacterial flora for their human host.

Normal Flora of the Oral Cavity

The presence of nutrients, epithelial debris, and secretions makes the mouth a favorable habitat for a great variety of bacteria. Oral bacteria include streptococci, lactobacilli, staphylococci and corynebacteria, with a great number of anaerobes, especially bacteroides.

The mouth presents a succession of different ecological situations with age, and this corresponds with changes in the composition of the normal flora. At birth, the oral cavity is composed solely of the soft tissues of the lips, cheeks, tongue and palate, which are kept moist by the secretions of the salivary glands. At birth the oral cavity is sterile but rapidly becomes colonized from the environment, particularly from the mother in the first feeding. *Streptococcus salivarius* is dominant and may make up 98% of the total oral flora until the appearance of the teeth (6–9 months in humans). The eruption of the teeth during the first year leads to colonization by *S. mutans* and *S. sanguis*. These bacteria require a nondesquamating (nonepithelial) surface in order to colonize. They

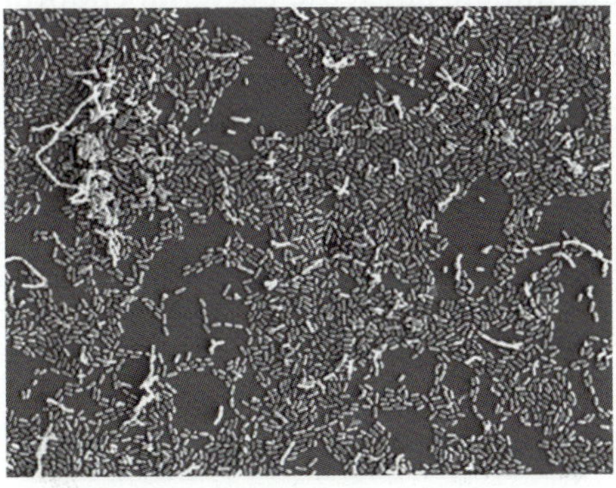

Various streptococci in a biofilm in the oral cavity

will persist as long as teeth remain. Other strains of streptococci adhere strongly to the gums and cheeks but not to the teeth. The creation of the gingival crevice area (supporting structures of the teeth) increases the habitat for the variety of anaerobic species found. The complexity of the oral flora continues to increase with time bacteroides and spirochetes colonize around puberty.

The normal bacterial flora of the oral cavity clearly benefit from their host who provides nutrients and habitat. There may be benefits, as well, to the host. The normal flora occupy available colonization sites which makes it more difficult for other microorganisms (nonindigenous species) to become established. Also, the oral flora contribute to host nutrition through the synthesis of vitamins, and they contribute to immunity by inducing low levels of circulating and secretory antibodies that may cross react with pathogens. Finally, the oral bacteria exert microbial antagonism against nonindigenous species by production of inhibitory substances such as fatty acids, peroxides and bacteriocins.

On the other hand, the oral flora are the usual cause of various oral diseases in humans, including abscesses, dental caries, gingivitis, and periodontal disease. If oral bacteria can gain entrance into deeper tissues, they may cause abscesses of alveolar bone, lung, brain, or the extremities. Such infections usually contain mixtures of bacteria with *Bacteroides melaninogenicus* often playing a dominant role. If oral streptococci are introduced into wounds created by dental manipulation or treatment, they may adhere to heart valves and initiate subacute bacterial endocarditis.

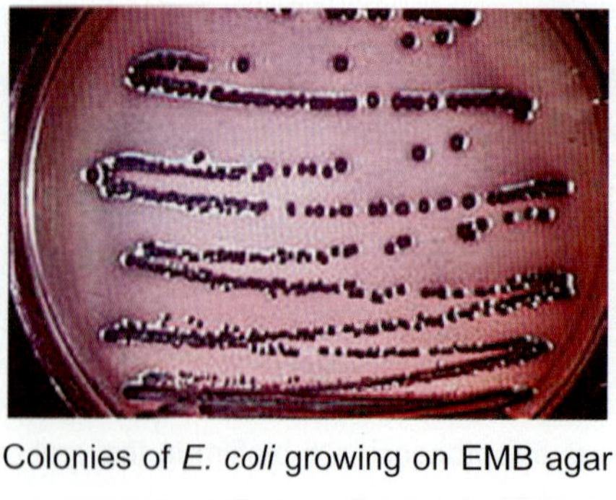

Colonies of *E. coli* growing on EMB agar

Normal Flora of the Gastrointestinal Tract

The bacterial flora of the gastrointestinal (GI) tract of animals has been studied more extensively than that of any other site. The composition differs between various animal species, and within an animal species. In humans, there are differences in the composition of the flora which are influenced by age, diet, cultural conditions, and the use of antibiotics. The latter greatly perturbs the composition of the intestinal flora.

In the upper GI tract of adult humans, the esophagus contains only the bacteria swallowed with saliva and food. Because of the high acidity of the gastric juice, very few bacteria (mainly acid-tolerant lactobacilli) can be cultured from the normal stomach. However, at least half the population in the United States is colonized by a pathogenic bacterium, *Helicobacter pylori.* Since the 1980s, this bacterium has been known to be the cause of gastric ulcers, and it is probably a cause of gastric and duodenal cancer as well. The Australian microbiologist, Barry Marshall, received the Nobel Prize in Physiology and Medicine in 2005, for demonstrating the relationship between *Helicobacter* and gastric ulcers.

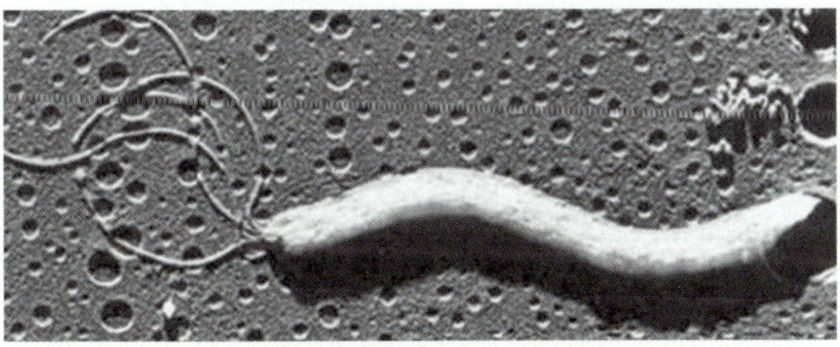

Helicobacter pylori. ASM

The proximal small intestine has a relatively sparse gram-positive flora, consisting mainly of lactobacilli and *Enterococcus faecalis.* This region has about 10^5–10^7 bacteria per ml of fluid. The distal part of the small intestine contains greater numbers of bacteria (10^8/ml) and additional species, including coliforms (*E. coli* and relatives) and *Bacteroides,* in addition to lactobacilli and enterococci.

The flora of the large intestine (colon) is qualitatively similar to that found in feces. Populations of bacteria in the colon reach levels of 10^{11}/ml feces. Coliforms become more prominent, and enterococci, clostridia and lactobacilli can be regularly found, but the predominant species are anaerobic *Bacteroides* and anaerobic lactic acid bacteria in the genus *Bifidobacterium* (*Bifidobacterium bifidum*). These organisms may outnumber *E. coli* by 1,000:1 to 10,000:1. Sometimes, significant numbers of anaerobic methanogens (up to 10^{10}/gm) may reside in the colon of humans. This is our only direct association with archaea as normal flora. The range of incidence of certain bacteria in the large intestine of humans is shown in Table 1.4.

Table 1.4: Bacteria found in the large intestine of humans.

Bacterium	*Range of incidence*
Bacteroides fragilis	100
Bacteroides melaninogenicus	100
Bacteroides oralis	100
Lactobacillus	20–60
Clostridium perfringens	25–35
Clostridium septicum	5–25
Clostridium tetani	1–35
Bifidobacterium bifidum	30–70
Staphylococcus aureus	30–50
Enterococcus faecalis	100
Escherichia coli	100
Salmonella enteritidis	3–7
Klebsiella sp.	40–80
Enterobacter sp.	40–80
Proteus mirabilis	5–55
Pseudomonas aeruginosa	3–11
Peptostreptococcus sp.	Common
Peptococcus sp.	Common

At birth the entire intestinal tract is sterile, but bacteria enter with the first feed. The initial colonizing bacteria vary with the food source of the infant. In breast-fed infants, bifidobacteria account for more than 90% of the total intestinal bacteria. *Enterobacteriaceae* and enterococci are regularly present, but in low proportions, while bacteroides, staphylococci, lactobacilli and clostridia are practically absent. In bottle-fed infants, bifidobacteria are not predominant. When breast-fed infants are switched to a diet of cow's milk or solid food, bifidobacteria are progressively joined by enterics, bacteroides, enterococci lactobacilli and clostridia. Apparently, human milk contains a growth factor that enriches for growth of bifidobacteria, and these bacteria play an important role in preventing colonization of the infant intestinal tract by non indigenous or pathogenic species.

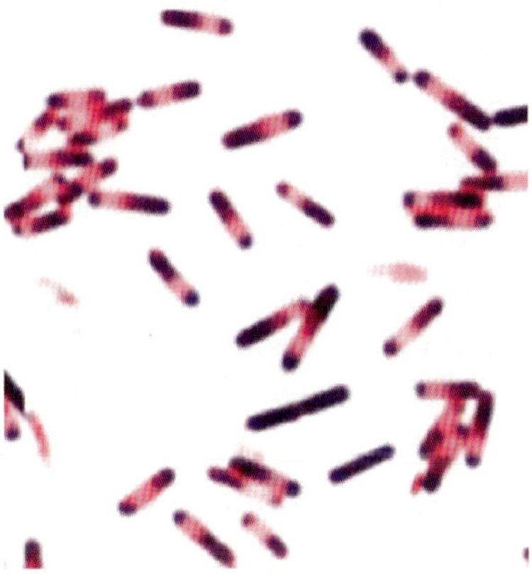

Clostridium difficile. Gram stain. The growth of "C. difficile" in the intestinal tract is normally held in check by other members of the normal flora. When antibiotics given for other infections cause collateral damage to the normal intestinal flora, the clostridium may be able to "grow out" and produce a serious diarrheal syndrome called pseudomembranous colitis. This is an example of an "antibiotic induced diarrheal disease"

The composition of the flora of the gastrointestinal tract varies along the tract (at longitudinal levels) and across the tract (at horizontal levels) where certain bacteria attach to the gastrointestinal epithelium and others occur in the lumen. There is frequently a very close association between specific bacteria in the intestinal ecosystem and specific gut tissues or cells (evidence of tissue tropism and specific adherence). Gram-positive bacteria, such as the streptococci and lactobacilli, are thought to adhere to the gastrointestinal epithelium using polysaccharide capsules or cell wall teichoic acids to attach to specific receptors on the epithelial cells. Gram-negative bacteria such as the enterics may attach by means of specific fimbriae which bind to glycoproteins on the epithelial cell surface.

It is in the intestinal tract that we see the greatest effect of the bacterial flora on their host. This is due to their large mass and numbers. Bacteria in the human GI tract have been shown to produce vitamins and may otherwise contribute to nutrition and digestion. But their most important effects are in their ability to protect their host from establishment and infection by alien microbes and their ability to stimulate the development and the activity of the immunological tissues.

On the other hand, some of the bacteria in the colon (e.g. *Bacteroides*) have been shown to produce metabolites that are carcinogenic, and there may be an increased incidence of colon cancer associated with these bacteria. Alterations in the GI flora brought on by poor nutrition or perturbance with antibiotics can cause shift in population and colonization by nonresidents that leads to gastrointestinal disease.

Beneficial Effects of the Normal Flora

The effects of the normal flora are inferred by microbiologists from experimental comparisons between **"germ-free" animals** (which are not colonized by any microbes) and conventional animals (which are colonized with a typical normal flora). Briefly, some of the characteristics of a germ-free animals that are thought to be due to lack of exposure to a normal flora are:

1. Vitamin deficiencies, especially vitamin K and vitamin B_{12}
2. Increased susceptibility to infectious disease
3. Poorly developed immune system, especially in the gastrointestinal tract
4. Lack of "natural antibody" or natural immunity to bacterial infection.

Because these conditions in germ-free mice and hamsters do not occur in conventional animals, or are alleviated by introduction of a bacterial flora (at the appropriate time of development), it is tempting to conclude that the human normal flora make similar contributions to human nutrition, health and development. The overall beneficial effects of microbes are summarized below.

1. **The normal flora synthesize and excrete vitamins** in excess of their own needs, which can be absorbed as nutrients by their host. For example, in humans, enteric bacteria secrete vitamin K and Vitamin B_{12}, and lactic acid bacteria produce certain B-vitamins. Germ-free animals may be deficient in vitamin K to the extent that it is necessary to supplement their diets.
2. **The normal flora prevent colonization by pathogens** by competing for attachment sites or for essential nutrients. This is thought to be their most important beneficial effect, which has been demonstrated in the oral cavity, the intestine, the skin, and the vaginal epithelium. In some experiments, germ-free animals can be infected by 10 *Salmonella* bacteria, while the infectious dose for conventional animals is near 10^6 cells.
3. **The normal flora may antagonize other bacteria** through the production of substances which inhibit or kill nonindigenous species. The intestinal bacteria produce a variety of substances ranging from relatively nonspecific fatty acids and peroxides to highly specific bacteriocins, which inhibit or kill other bacteria.
4. **The normal flora stimulate the development of certain tissues**, i.e. the cecum and certain lymphatic tissues (Peyer's patch) in the GI tract. The cecum of germ-free animals is enlarged, thin-walled, and fluid-filled, compared to that organ in conventional animals. Also, based on the ability to undergo immunological stimulation, the intestinal lymphatic tissues of germ-free animals are poorly-developed compared to conventional animals.
5. **The normal flora stimulate the production of natural antibodies.** Since the normal flora behave as antigens in an animal, they induce an immunological response, in particular, an antibody-mediated immune (AMI) response. Low levels of antibodies produced against components of the normal flora are known to cross react with certain related pathogens, and thereby prevent infection or invasion. Antibodies produced against antigenic components of the normal flora are sometimes referred to as "natural" antibodies, and such antibodies are lacking in germ-free animals.

Harmful Effects of the Normal Flora

Harmful effects of the normal flora, some of which are observed in studies with germ-free animals, can be put in the following categories. All but the last two are fairly insignificant.

1. **Bacterial synergism** between a member of the normal flora and a potential pathogen. This means that one organism is helping another to grow or survive. There are examples of a member of the normal flora supplying a vitamin or some other growth factor that a pathogen needs in order to grow. This is called **cross-feeding** between microbes. Another example of synergism occurs during treatment of **"staph-protected infections"** when a penicillin-resistant staphylococcus that is a component of the normal flora shares its drug resistance with pathogens that are otherwise susceptible to the drug.
2. **Competition for nutrients:** Bacteria in the gastrointestinal tract must absorb some of the host's nutrients for their own needs. However, in general, they transform them into other metabolisable compounds, but some nutrient(s) may be lost to the host. Germ-free animals are known to grow more rapidly and efficiently than conventional animals. One explanation for incorporating antibiotics into the food of swine, cows and poultry is that the animal grows faster and can therefore be marketed earlier. Unfortunately, this practice contributes to the development and spread of bacterial antibiotic resistance within the farm animals, as well as humans.
3. **Induction of a low-grade toxemia:** Minute amounts of bacterial toxins (e.g. endotoxin) may be found in the circulation. Of course, it is these small amounts of bacterial antigen that stimulate the formation of natural antibodies.
4. **The normal flora may be agents of disease.** Members of the normal flora may cause **endogenous disease** if they reach a site or tissue where they cannot be restricted or tolerated by the host defenses. Many of the normal flora are potential pathogens, and if they gain access to a compromised tissue from which they can invade, disease may result.
5. **Transfer to susceptible hosts.** Some pathogens of humans that are members of the normal flora may also rely on their host for transfer to other individuals where they can produce disease.

This includes the pathogens that colonize the upper respiratory tract such as *Neisseria meningitidis, Streptococcus pneumoniae, Haemophilus influenzae* and *Staphylococcus aureus,* and potential pathogens such as *E. coli, Salmonella* or *Clostridium* in the gastrointestinal tract.

Dental Caries, Gingivitis and Periodontal Disease

The most frequent and economically-important condition in humans resulting from interactions with our normal flora is probably dental caries. Dental plaque, dental caries, gingivitis and periodontal disease result from actions initiated and carried out by the normal bacterial flora.

Dental plaque, which is material adhering to the teeth, consists of bacterial cells (60–70% the volume of the plaque), salivary polymers, and bacterial extracellular products. Plaque is a naturally-constructed biofilm, in which the consortia of bacteria may reach a thickness of 300–500 cells on the surfaces of the teeth. These accumulations subject the teeth and gingival tissues to high concentrations of bacterial metabolites, which result in dental disease.

The dominant bacterial species in dental plaque are *Streptococcus sanguis* and *Streptococcus mutans,* both of which are considered responsible for plaque.

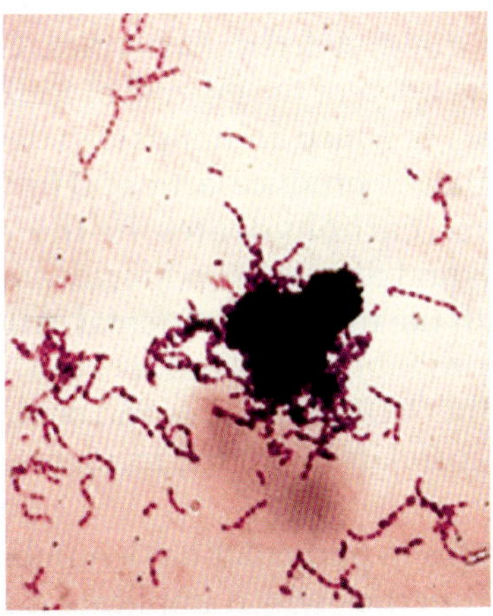

Streptococcus mutans. Gram stain. CDC

Plaque formation is initiated by a weak attachment of the streptococcal cells to salivary glycoproteins forming a pellicle on the surface of the teeth. This is followed by a stronger attachment by means of extracellular sticky polymers of glucose (glucans) which are synthesized by the bacteria from dietary sugars (principally sucrose). An enzyme on the cell surface of *Streptococcus mutans*, glycosyl transferase, is involved in initial attachment of the bacterial cells to the tooth surface and in the conversion of sucrose to dextran polymers (glucans) which form plaque.

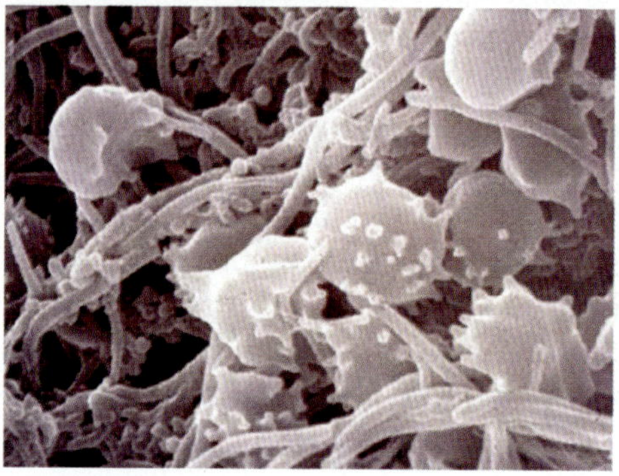

Dental plaque, scanning electron micrograph illustrating the diversity of microbes in plaque

Dental caries is the destruction of the enamel, dentin or cementum of teeth due to bacterial activities. Caries are initiated by direct demineralization of the enamel of teeth due to lactic acid and other organic acids which accumulate in dental plaque. Lactic acid bacteria in the plaque produce lactic acid from the fermentation of sugar and other carbohydrates in the diet of the host. *Streptococcus mutans* and *Streptococcus sanguis* are most consistently been associated with the initiation of dental caries, but other lactic acid bacteria are probably involved as well. These organisms normally colonize the occlusal fissures and contact points between the teeth, and this correlates with the incidence of decay on these surfaces.

Streptococcus mutans in particular has a number of physiological and biochemical properties which implicate it in the initiation of dental caries.

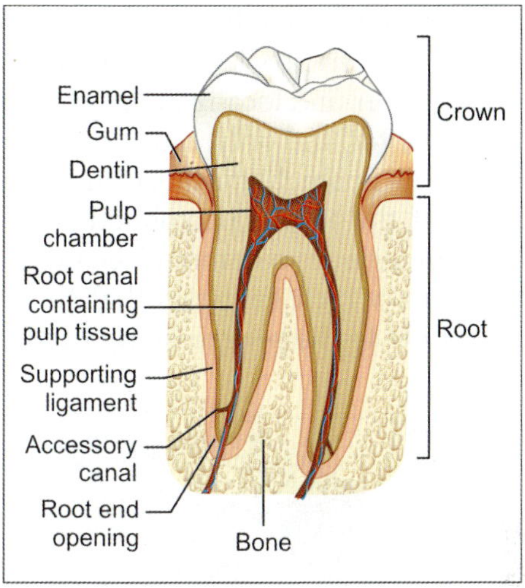

Cross-section of a tooth illustrating the various structural regions susceptible to colonization or attack by microbes

1. It is a regular component of the normal oral flora of humans which occurs in relatively large numbers. It readily colonizes tooth surfaces: Salivary components (mucins, which are glycoproteins) form a thin film on the tooth called the enamel pellicle. The adsorbed mucins are thought to serve as molecular receptors for ligands on the bacterial cell surface.
2. It contains a cell-bound protein, glycosyl transferase, that serves an adhesin for attachment to the tooth, and as an enzyme that polymerizes dietary sugar into glucans that leads to the formation of plaque.
3. It produces lactic acid from the utilization of dietary carbohydrate which demineralizes tooth enamel. *S. mutans* produces more lactic acid and is more acid-tolerant than most other streptococci.
4. It stores polysaccharides made from dietary sugar which can be utilized as reserve carbon and energy sources for production of lactic acid. The extracellular glucans formed by *S. mutans* are, in fact, bacterial capsular polysaccharides that function as carbohydrate reserves. The organisms can also form intracellular polysaccharides from sugar which are stored in cells and then metabolized to lactic acid.

Streptococcus mutans appears to be important in the initiation of dental caries because its activities lead to colonization of the tooth surfaces, plaque formation, and localized demineralization of tooth enamel. It is not, however, the only cause of dental decay. After initial weakening of the enamel, various oral bacteria gain access to interior regions of the tooth. Lactobacilli, *Actinomyces,* and various proteolytic bacteria are commonly found in human carious dentin and cementum, which suggests that they are secondary invaders that contribute to the progression of the lesions.

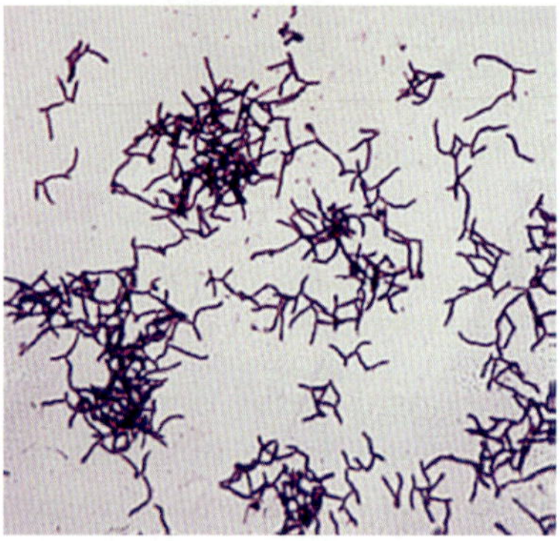

Actinomyces israelii

Periodontal diseases are bacterial infections that affect the supporting structures of the teeth (gingiva, cementum, periodontal membrane and alveolar bone). The most common form, **gingivitis**, is an inflammatory condition of the gums. It is associated with accumulations of bacterial plaque in the area. Increased populations of *Actinomyces* have been found.

Diseases that are confined to the gum usually do not lead to loss of teeth, but there are other more serious forms of periodontal disease that affect periodontal membrane and alveolar bone resulting in tooth loss. Bacteria in these lesions are very complex populations consisting of Gram-positive organisms (including *Actinomyces* and streptococci) and Gram-negative organisms (including spirochetes and *Bacteroides*). The mechanisms of tissue destruction in periodontal disease are not clearly defined but hydrolytic enzymes, endotoxins, and other toxic bacterial metabolites seem to be involved.

Factors Influencing the Oral Ecosystem

The growth of oral microorganisms is influenced by a variety of factors such as temperature, pH, oxidation-reduction potential, the availability of nutrients and water, the anatomy of the oral structures, salivary flow, and antimicrobial substances. Each factor in a given oral habitat influences the selection of oral microorganisms and helps maintain the equilibrium among bacterial populations. The result of these selective pressures has already been observed in the differences in the oral microbiota among the different sites of the oral cavity.

The formation of dental plaque on smooth surfaces has been widely studied *in vitro* and *in vivo* and represents a good example of the force involved to maintain the homeostasis of the oral ecosystems. The development of dental plaque follows a general bacterial succession pattern under the control of several factors (reviewed in reference). After tooth brushing, dental plaque is initiated by the deposition of an acellular proteinaceous film, termed the acquired pellicle. The major constituents of the pellicle are components of saliva and gingival crevicular fluid such as proteins (albumin, lysozyme, proline-rich proteins), glycoproteins (lactoferrin, IgA, IgG, amylase), phosphoproteins, and lipids. Bacterial components such as glucosyltransferase are also present. The bacteria colonize the pellicle within the first 2 to 4 hours after cleaning. The pioneer microorganisms consist mainly of streptococci (*S. sanguis*, *S. oralis*, and *S. mitis*) and, in smaller numbers, *Neisseria* and *Actinomyces*. The bacteria with low affinity for the pellicle are eliminated by salivary flow. After initial colonization, the pioneer microorganisms grow rapidly, forming microcolonies that are embedded in an extracellular matrix composed of bacterial and host molecules. During this process, the alteration of the environment by pioneers allows the colonization by other bacterial groups such as *Veillonella* and *Haemophilus* (48 h after tooth brushing). Several bacterial interrelationships including coaggregation, production of antibacterial substances, and food chains contribute to increase the diversity of the bacterial community. Also, the consumption of oxygen by aerobic species favors the colonization of obligately anaerobic microorganisms such as *Fusobacterium*, *Bacteroides*, and spirochetes (1 to 2 weeks). If the plaque is left to accumulate, the complexity of the microflora increases until a climax community has been established (2 to 3 weeks). As described above, an imbalance in the plaque ecosystem may lead to the development of oral diseases. For example, the uncontrolled development of supragingival plaque in the absence of oral hygiene may cause gingivitis and subsequent colonization of the subgingival crevice by periodontopathogens. A high sucrose intake may lead to a higher colonization of plaque by *S. mutans* and *Lactobacillus* and to the development of caries.

In the following sections, each factor that may influence the physicochemical environment and the colonization of the oral cavity will be described in detail. All these factors are interrelated and depend on host and microbial activities as well as external factors such as diet or oral hygiene. For better clarity, these factors have been divided into four categories: Physicochemical, host related, microbial, and external.

Physicochemical Factors

The physicochemical factors result from the combined action of host, microbial, and external factors. In all *in vivo* and *in vitro* systems, the growth of microorganisms is influenced by five important variables: Temperature, pH, availability of water, availability of nutrients, and oxidation-reduction potential. As the mouth is constantly bathed by saliva and crevicular fluid, water is not considered to be a limiting factor.

Temperature

The temperature in the oral cavity is relatively constant (34 to 36°C), which allows a wide range of microorganisms to grow. The temperature may be more variable on the mucosal and tooth supragingival surface. During food intake, microorganisms colonizing these sites are exposed to hot and cold meals and probably must adapt to these extreme variations of temperature. However, to our knowledge, no data are available on the effect of this short period of temperature variation on the metabolism of oral bacteria.

pH

The pH or hydrogen ion concentration of an environment affects microorganisms and microbial enzymes directly and also influences the dissolution of many molecules that indirectly influence microorganisms. Microorganisms generally cannot tolerate extreme pH values. In the oral cavity, the pH is maintained near neutrality (6.7 to 7.3) by saliva. The saliva contributes to maintenance of the pH by two mechanisms. First, the flow of saliva eliminates carbohydrates that could be metabolized by bacteria and removes acids produced by bacteria. Second, acidity from drinks and foods, as well as from bacterial activity, is neutralized by the buffering activity of saliva. Bicarbonate is the major salivary buffering system of saliva, but peptides, proteins, and phosphates are also involved. Increases in pH also result from bacteria that metabolize sialine and urea into ammonia. Acids that are produced by the microbial metabolism of carbohydrates may accumulate in dental plaque because of the slow diffusion of saliva through dental plaque. Following sugar intake, the pH of dental plaque may decrease to below 5.0.

The pH is an important parameter in oral microbial ecology. Frequent sugar intake favours the growth of aciduric bacteria such as *Lactobacillus* and *S. mutans* and predisposes to caries formation. An increased colonization by *S. mutans* was demonstrated by simply rinsing the mouth with low-pH buffers. *In vitro* studies have also shown that gradual decreases in pH in glucose-pulsed cultures favored *S. mutans* and *Lactobacillus* while populations of *S. sanguis, S. mitior, P. intermedia,* and *F. nucleatum* were reduced. When the pH of the chemostat was controlled at 7.0, the glucose pulse had little effect on the microbial populations, suggesting that the low pH generated from carbohydrate metabolism, rather than carbohydrate availability per se, is responsible for the shift in composition of the oral microbiota *in vivo*.

The subgingival area is bathed by gingival fluid and is not controlled by the buffering salivary activity. The pH in the gingival crevice may vary between 7.5 and 8.5, while the crevicular fluid ranges from pH 7.5 to 7.9. An alkaline pH in gingival crevices and periodontal pockets may exert a selective force towards the colonization of periodontopathogens.

Oxidation – Reduction Potential and Anaerobiosis

Many enzymatic reactions are oxidation – reduction reactions in which one compound is oxidized and another compound is reduced. The proportion of oxidized to reduced components constitutes the oxidation – reduction potential or redox potential (E_h). The E_h is greatly influenced by the presence or absence of molecular oxygen, which is the most common electron acceptor. Anaerobic bacteria need a reducing environment (negative E_h) for growth, while aerobic bacteria need an oxidizing environment (positive E_h). The mouth is characterized by a wide range of oxidation-reduction potentials, allowing the growth of aerobic, facultative anaerobic, and anaerobic bacteria. In general, the dorsum of the tongue and the buccal and palatal mucosa are aerobic environments with positive E_h, thus better supporting the growth of facultative anaerobic bacteria. The gingival crevice and the approximal surfaces of the teeth (surfaces between teeth) possess the lowest E_h and the highest concentration of obligately anaerobic bacteria. The E_h values vary between +158 and +542 mV in saliva but may reach 300 mV in gingival crevices. The E_h also varies during plaque formation, changing from positive values (+294 mV) on clean tooth surfaces to negative values (141 mV) after 7 days. The fall in E_h during plaque formation is the result of oxygen consumption by facultative anaerobic bacteria as well as a reduction in the ability of oxygen to diffuse through the plaque. This explains in part the increased in number of obligately anaerobic bacteria during plaque formation.

Nutrients

Chemostat studies and a study with mice suggest that the levels of most bacterial populations are strongly controlled by substrate availability. Each bacterial species must be more efficient than the rest in utilizing one or a few particular substrates under certain conditions. According to Liebig's law of the minimum, the growth of each organism is limited by the required substrate that is present in the lowest concentration. In the oral cavity, microorganisms living in the supragingival environment have access to nutrients from both endogenous (saliva) and exogenous (diet) origin. Saliva is an important source of nutrients and can sustain normal growth of microorganisms in the absence of exogenous nutrients. Saliva contains water, carbohydrates, glycoproteins, proteins, amino acids, gases, and several ions including sodium, potassium, calcium, chloride, bicarbonate, and phosphate. Among exogenous dietary components, carbohydrates and proteins have the greatest influence on the composition of the oral microbiota.

The gingival crevice is not exposed to dietary components and saliva, and its principal source of nutrients is the gingival crevicular fluid. The crevicular fluid originates from plasma and is an excellent source of nutrients for fastidious microorganisms. It contains growth factors such as hemin and vitamin K required by *P. gingivalis*, a gram-negative rod associated with adult periodontitis.

Many nutritional interrelationships also occur between microorganisms. Some microorganisms cooperate for the degradation of nutrients. Some bacteria also use nutrients and other substances produced by other microorganisms.

Host Factors

Host Defense Mechanisms

The supragingival environment of the oral cavity is controlled primarily by saliva. The continuous flow of saliva increased by the muscular activity of the lips and tongue removes a large number of bacteria from teeth and mucosal surfaces. Saliva also contains several specific and nonspecific defense factors. SIgA is the principal specific defense factor of saliva, and its role is discussed more extensively below. The nonspecific defense factors include mucins, nonimmune salivary glycoproteins, lactoferrin, lysozyme, peroxidase, histatins, and cystatins.

Mucins are high-molecular-weight glycoproteins produced by submandibular, sublingual, and numerous minor salivary glands. They are the principal organic constituent of mucus, the slimy viscoelastic material that envelopes all mucosal surfaces of the body. Saliva contains two forms of mucins, MG1 and MG2. The MG1 mucin, which has a molecular mass greater than 1,000 kDa, is involved mainly in tissue coating; MG2, which has a molecular mass of 125 kDa, affects the aggregation and adherence of streptococci. In the oral cavity, mucins provide a protective coating for both soft and hard tissues. The mucins form a viscous slime layer on oral mucosa that traps microorganisms and antigens, limiting their penetration into the tissues. Potentially harmful microorganisms are thus eliminated by the continuous renewal of the mucous layer combined with the washing action of salivary flow. Mucins are also constituents of the acquired pellicle and may protect teeth from acid demineralization.

The role of mucins and other nonimmune salivary glycoproteins in bacterial adherence is complex. When salivary glycoproteins are adsorbed on solid surfaces, they may bind to bacteria and promote bacterial adherence. Conversely, some of these glycoproteins, when free in saliva, may prevent bacterial colonization by binding to their adhesins or by agglutinating bacteria in saliva. This type of aggregation may facilitate the removal of oral bacteria by swallowing. Saliva from subjects with low levels of mutans streptococci aggregate these bacteria more efficiently, suggesting a protective role for salivary agglutinins. In vitro, pretreatment of *S. sanguis* and *S. mutans* with salivary glycoproteins prevented their attachment to hydroxyapatite. The phenomena of adherence and aggregation may be mediated by two different binding mechanisms. Saliva agglutinin (300 kDa), which acts as a receptor for antigen I/II of *S. mutans*, mediates both aggregation and adherence. However, the interaction involves different regions of the antigen I/II and different binding specificities.

Saliva also possesses defense factors with direct antimicrobial activity in vitro. A group of salivary proteins, lysozyme, lactoferrin, and peroxidase, act in conjunction with other components of saliva to limit the growth of bacteria or kill them directly.

Lysozyme is a small cationic protein that is present in all major body fluids; it is secreted by intercalated duct cells. Lysozyme can lyse some bacterial species by hydrolyzing glycosidic linkages in the cell wall peptidoglycan. It may also cause lysis of bacterial cells by interacting with monovalent anions, such as thiocyanate, perchlorate, iodide, bromide, bicarbonate, nitrate, and fluoride, and with proteases found in saliva. The combination leads to destabilization of the cell membrane probably by activation and deregulation of endogenous bacterial autolysins. In vitro, the bacteriolytic activity of the lysozyme-protease-monovalent anion system has been demonstrated against *S. mutans*, *L. casei*, and *F. nucleatum*. In addition, lysozyme can aggregate oral bacterial cells and inhibit their colonization on mucosal surfaces and teeth. In vivo, an inverse correlation has been found between the concentration of lysozyme and the accumulation of dental plaque.

Lactoferrin is an iron-binding glycoprotein produced by intercalated duct cells. It inhibits microbial growth, probably by sequestering iron in the environment. In addition, iron-free lactoferrin (apolactoferrin) possesses a direct, iron-independent bactericidal effect against various oral bacterial strains including *S. mutans*. Apolactoferrin was shown to agglutinate *S. mutans* but not other species of streptococci, *P. gingivalis*, or *A. actinomycetemcomitans*. Recent results with bovine lactoferrin indicate a bactericidal activity of the N-terminal portion of the lactoferrin molecule.

Salivary peroxidase is an enzyme secreted by salivary gland acinar cells. It is part of an antimicrobial system that involves the oxidation of salivary thiocyanate to hypothiocyanite and hypothiocyanous acid by hydrogen peroxide, generated by oral bacteria. These oxidizing agents react with sulfhydryl groups of the enzymes involved in glycolysis and sugar transport. Salivary peroxidase removes toxic hydrogen peroxide produced by oral microorganisms and can reduce acid production in dental plaque. Salivary peroxidase has been shown to retain activity when adsorbed on hydroxyapatite and so should be effective at the enamel-plaque interface. The activation of peroxidase systems in vivo by addition of appropriate amounts of exogenous hydrogen peroxide reduces plaque accumulation, gingivitis and caries.

Histatins (histidine-rich peptides) are a family of small basic peptides (3 to 5 kDa), with a high content of histidine, that are produced by acinar cells. They inhibit the development of *Candida albicans* from the noninfective to the infective form. They can also inhibit coaggregation between *P. gingivalis* and *S. mitis*, aggregate oral streptococci, and inhibit the growth of *S. mutans*.

Cystatins are a family of cysteine-containing phosphoproteins that are secreted by acinar cells. These proteins are also present in plasma and may reach the oral cavity via the gingival crevicular fluid. Cystatins act mainly as thiol protease inhibitors and can inhibit proteases produced by suspected periodontopathogens.

Saliva does not gain access to the gingival crevice, and this area of the oral cavity is almost essentially controlled by the antimicrobial factors of plasma. Cellular and humoral components of blood can reach the gingival crevice of the oral cavity by the flow of gingival fluid through the junctional epithelium. Even in the healthy state, there is a continuous flow of small quantities of fluid and leukocytes from the gingival capillaries through the crevicular epithelium into the gingival crevice. This flow increases greatly with inflammation induced by plaque accumulation. The continuous flow of gingival fluid from the crevice to the oral cavity removes nonadherent bacterial cells. Gingival fluid also contains antimicrobial substances including IgM, IgG, IgA, complement, and leukocytes. These factors are primarily protective against microbial invasion, but, as seen above, the inflammation may become destructive, resulting in loss of periodontal attachment.

The leukocytes in gingival crevicular fluid are composed of 90% polymorphonuclear leukocytes (PMNs) and 10% mononuclear cells. Among the mononuclear cells, 60% are B lymphocytes, 20 to 30% are T lymphocytes, and 10 to 15% are macrophages. About 80% of PMNs are viable and functional within the crevice. The cells are capable of phagocytosis and of killing microorganisms, although the efficiency of phagocytosis is reduced compared with that of blood neutrophils. The PMNs perhaps remain functional at a short distance from the gingival margin by the flow of gingival fluid along the tooth surface, but once neutrophils are in saliva, they degenerate, due to osmotic lysis. Lysozyme and peroxidase that are released from the lysosome of PMNs during phagocytosis might also control microbial growth in the gingival crevice.

Components of the complement cascade are present in the gingival crevicular fluid. In subjects with healthy gingivae, C3 and C4 components of complement can be detected. During gingival inflammation, C3a, C3b, and C5a appear, suggesting that complement activation may have occurred *in vivo*. Complement factors may initiate bacterial cell lysis or enhance phagocytosis of microorganisms.

The IgG, IgM, and IgA antibodies directed against a variety of oral microorganisms have been detected in plasma and crevicular fluid even in healthy individuals. These antibodies may influence the oral microbiota by interfering with adherence or by inhibiting bacterial metabolism. Furthermore, the IgG antibodies may enhance phagocytosis and killing of oral microorganisms through activation of complement or opsonization. It has been demonstrated that systemic immunization of animals with periodontopathogens may reduce the colonization of these bacteria in the gingival crevice and reduce periodontal destruction. However, since periodontal diseases are of multifactorial origin, systemic immunization with periodontopathogens may also enhance the destruction of alveolar bone. The immune response itself may contribute significantly to the periodontal destruction, sometimes even more than the pathogens.

Some indications suggest that serum antibodies may also regulate the bacterial colonization of supragingival surfaces of teeth. Although conflicting evidence exists, some studies have reported an inverse correlation between the level of serum IgG antibodies against *S. mutans* and the level of this bacterium in dental plaque or the frequency of caries. Also, systemic immunization of nonhuman primates with *S. mutans* antigens led to reduced levels of *S. mutans* and less caries.

All specific and nonspecific antimicrobial factors in the oral cavity do not act in isolation. Synergistic and antagonistic interactions among antimicrobial factors may influence their actions. Mucins serve to concentrate other antimicrobial substances, including lysozyme, IgA, and cystatins, at the mucosal surface. SIgA enhances the antimicrobial activity of lactoferrin, salivary peroxidase, agglutinins, and mucins. Similarly, the polycationic antimembrane effect of lysozyme may be enhanced by salivary peroxidase and histatins. In contrast, salivary peroxidase may block the bactericidal effect of lactoferrin.

Age

The composition of the oral microbiota varies with the age of the host. Age-related changes in the oral microflora include those due to teeth eruption, changes in dietary habits, hormones, salivary flow, the immune system, or other factors.

The human oral cavity is usually sterile at birth. However, within 6 to 10 h after birth, microorganisms from the mother and to a lesser extent microorganisms from those present in the environment become established in the oral cavity. The pioneer species are usually streptococci, especially *S. mitis* biovar 1, *S. oralis*, and *S. salivarius*. During the first year of life, the oral microbiota contains *Streptococcus, Neisseria, Veillonella, Staphylococcus*, and, to a lesser degree, *Actinomyces, Lactobacillus, Rothia, Fusobacterium*, and *Prevotella*. Some species, such as *S. sanguis, S. mutans*, and *A. naeslundii* genospecies 2 (formerly *A. viscosus*), colonize the oral cavity only after tooth eruption. Following tooth eruption, the number and isolation frequency of obligately anaerobic bacteria increase. The number of black-pigmented anaerobes and spirochetes in the gingival crevice increases more extensively during adolescence, and this could be due to hormonal changes. Studies indicate that the next most important changes occur in the elderly and include an increased prevalence of staphylococci, lactobacilli, and *A. naeslundii* genospecies 2 (formerly *A. viscosus*) after the age of 70 years and an increase in the proportion of *Candida albicans* after 80 years. The change in the oral microbiota of elderly individuals were not related to denture wearing, medication, or disease but may be caused by a decrease in salivary flow, an impaired immune system, or nutritional deficiencies.

Staphylococci have been identified as the first colonizers, which are immediately followed by lactobacilli. The appearance of *Enterococcus faecalis* and members of the *Enterobacteriaceae* appeared to correspond to tooth eruption and the beginning of coprophagy. The proportion of *Lactobacillus murinus* sharply increased at weaning (20 days), probably due to changes in the composition and texture of the diet, from maternal milk to solid food. After the weaning period, no other significant changes were observed and the oral microbiota appeared to be completely stabilized when the mice reached the age of 6 to 8 weeks.

Hormonal Changes

It is well known that in humans, puberty and pregnancy are characterized by increased levels of steroid hormones in plasma and subsequently in the crevicular fluid and saliva. It is also well documented that pregnancy and puberty are associated with an increase in gingival inflammation which is accompanied by an increase in gingival exudate. It has been proposed that the exacerbations in gingival inflammation may be due to hormone-induced alterations in the microbiota of the gingival crevice. Microorganisms in the subgingival area that use hormones as growth factors may be favored during the period of hormone increase associated with puberty and pregnancy. Several investigators have described a transient increase in the number of black-pigmented gram-negative anaerobic bacteria in the subgingival microbiota during puberty or pregnancy. Kornman and Loeshe reported an increased proportion of *Prevotella intermedia* in the subgingival microbiota of pregnant woman, corresponding to an increased levels of estrogens and progesterone in plasma. They also demonstrated *in vitro* that progesterone or estradiol can substitute for vitamin K as an essential growth factor for *P. intermedia*. In contrast, other studies were unable to find any changes in the subgingival microbiota during puberty and pregnancy.

In mice, we did not find any modifications in the oral microbiota during puberty or pregnancy. However, in contrast to humans, mice do not usually harbor obligately anaerobic gram-negative rods. Furthermore, since the periods of puberty and pregnancy in mice are short, the bacteria might not be exposed to the hormones long enough to induce modifications in the oral microbiota.

Stress

Host stress may be associated with changes in hormones, salivary flow, dietary habits, and immune response. Few studies on the effect of stress on the indigenous microbiota have been performed, and most data have been obtained from studies of the intestinal microbiota of rodents. Stress in rodent is considered to be related to stimulation of the hypothalamic-pituitary-adrenocortical axis, which can lead to an elevation in corticosterone concentration. Crowding, fighting, and husbandry changes are some of the factors that are known to induce an increase in cortisol levels in plasma in mice and might play an important role in animal stress. These factors affect the behavior of mice in terms of feeding, sexual habits, grooming, rearing, and biting. The composition of the intestinal microbiota of rodents may be altered by a variety of unrelated disturbances, such as changes in environmental temperature, crowding in cages, and fighting among animals. These changes in the intestinal microbiota include a reduction in numbers of lactobacilli as well as fusiform and segmented filamentous bacteria. In our studies, some evidence suggested that the oral microbiota of mice may also be influenced by stress. Various factors such as shipping to our animal facility, husbandry modifications, and low temperature (in nude mice) led to decreases in the proportions of *Lactobacillus murinus* among the total cultivable oral microbiota.

Genetic Factors

The genetic background appears to influence the susceptibility to caries and periodontal diseases. This could in part be because the host genetic factors select for a microbiota with varying potential for causing oral diseases. The selection of a certain microbiota by the host is dependent on inherited immune factors, physiology, metabolism, mucus composition, or receptor-ligand interactions. Malamud et al. produced evidence for the inheritability of agglutinin activity and parotid flow rate. Genetic factors also seem to influence the intestinal and oral microbiota. Moore et al reported that the composition of subgingival microbiotas of monozygotic twins (11 to 14 years of age) was more similar than that of dizygotic twins. In contrast, we found that the genetic background does not seem to be an important factor in the composition of the oral microbiota of mice and that its influence is probably masked by environmental and husbandry factors. On the other hand, the oral indigenous microbiota of SPF mice is simple and the effect of the genetic background is perhaps observed only when the host is exposed to a more complex microbiota.

Bacterial Factors

Adherence

To get established in the oral cavity, microorganisms must first adhere to teeth or to mucosal surfaces. Adherence is essential for providing resistance to the flow of saliva. Adherence is mediated by adhesins on the surface of bacteria and by receptors on the oral surface. Microbial adhesins consist of polysaccharides, lipoteichoic acids, glucosyltransferases, and carbohydrate-binding proteins (lectins). These adhesins are found as cell wall components or are associated with cell structures, such as fimbriae, fibrils or capsules. The receptors may be salivary components (mucins, glycoproteins, amylase, lysozyme, IgA, IgG, proline-rich proteins, and statherins) or bacterial components (glucosyltransferases and glucans) that are bound to oral surfaces. The adherence may result from nonspecific physicochemical interactions between the bacteria and the oral surfaces. For example, lipoteicoic acids on microbial surfaces may interact with negatively charged host components through calcium ions or through hydrogen or hydrophobic bonding. However, these interactions cannot alone explain the selective attachment of bacteria to various oral surfaces. It is believed that another mechanism accounts for this selective colonization, perhaps involving specific or stereochemical interactions between bacterial adhesins and host receptors. In this case, the same physicochemical forces intervene but now act between extremely small, highly localized, spatially well organized opposing molecular groups. It is probable that the bacteria first adhere by nonspecific interactions which are followed by stronger stereochemical interactions. A number of specific interactions have been identified in adhesion to human tooth surfaces (reviewed in reference). The stereochemical interactions involved in bacterial adhesion in the oral cavity are analogous to the interactions between antigen and antibody or between an enzyme and its substrate. For example, type 1 fimbriae of *Actinomyces naeslundii* genospecies 2 and surface antigen I/II (protein P1) of *S. mutans* can bind to proline-rich proteins, *Streptococcus gordonii* can bind to α-amylase, and *A. naeslundii* genospecies 2 and *Fusobacterium nucleatum* can interact with statherin. Another adherence strategy involves a lectin-like bacterial protein with the complementary carbohydrate receptor located on glycoproteins. The interaction may be inhibited in vitro by adding the specific carbohydrate. *S. sanguis* can bind to sialic acid-containing oligosaccharides of the low-molecular-weight salivary mucin (MG2). Type 2 fimbriae of *Actinomyces* binds to the beta-linked galactose glycoprotein on epithelial cell surfaces.

Bacteria may also colonize host surfaces by adhering to other bacteria, and several examples of coaggregation between human oral bacterial species have been demonstrated in vitro. *Streptococcus* spp. aggregates with *Actinomyces* spp., *F. nucleatum*, *Veillonella*, and *Haemophilus parainfluenzae. F. nucleatum* binds with *A. actinomytemcomitans*, *Porphyromonas gingivalis*, *H. parainfluenzae*, and *Treponema* spp. Most coaggregates that have been studied in detail involve two strains from different genera; this is referred to as intergeneric coaggregation. Intrageneric coaggregation is seen almost exclusively within oral viridans streptococci. Many of these interactions appear to be mediated by a lectin from one cell type that interacts with a complementary carbohydrate receptor from the other cell type. Coaggregation may be important in the development of dental plaque because it allows the colonization of bacteria that are not able to adhere directly to the acquired pellicle. However, almost all the data on coaggregation derive from in vitro experiments. Indications that the phenomenon exist in vivo are the microscopic observations of dental plaque revealing the presence of two types of structures: Gram-positive filaments covered by gram-positive cocci, referred to as corn cobs; and large filaments surrounded by gram-negative rods or short filaments, referred to as bristle brushes.

Another example of coaggregation is the synthesis of extracellular polysaccharides from sucrose by mutans streptococci. The glucosyltransferases that are bound to the surface of mutans streptococci synthesize glucans in the presence of sucrose. Thus, the polymers are cell associated and can bind to the tooth surface or to other bacteria via other glucosyltransferases or via independent glucan-binding components. These polysaccharides consolidate bacterial attachment to teeth and contribute to an increased stability of the plaque matrix. The synthesis of polymers by mutans streptococci in the presence of sucrose is probably one of the factors implicated in caries formation.

Bacterial Interactions

A variety of beneficial and antagonistic interactions may help in maintaining the homeostasis of the oral microbiota. Most of these bacterial interrelationship have been characterized *in vitro* or in animal models, and it is assumed that they operate in the same way in the human oral cavity.

Coaggregation is one example of commensalism and synergism that occurs between microbial species. Coaggregation allows the indirect adherence of some bacteria on oral surfaces. In addition, it was demonstrated that coaggregated cells were more resistant to phagocytosis and killing by neutrophils *in vitro* and *in vivo*. Several other examples of positive interactions are likely to occur in the oral cavity. The utilization of oxygen by facultative anaerobic bacteria reduces the oxygen concentration and the E_h to levels that allow the colonization of anaerobic bacteria. Different bacterial species may also cooperate in the utilization of substrates that they could not metabolize alone. In laboratory studies, *P. gingivalis* and *F. nucleatum* were shown to hydrolyze casein synergistically. Chemostat studies indicated that glycoprotein degradation may involve the synergistic action of several species possessing complementary patterns of glycosidase and protease activities. The development of complex food chains also contributes to the diversity and stability of oral ecosystems. For example, the metabolism of carbohydrates by *Streptococcus* and *Actinomyces* generates lactate, which may be used by *Veillonella*. The utilization of lactic acid by *Veillonella* produces vitamin K, required by black-pigmented gram-negative rods, and H_2, used by *Wolinella.*

Competition and antagonism mechanisms among oral resident bacteria may help to maintain the ecological balance by preventing the overgrowth of some resident bacterial species or the establishment of allochthonous bacteria. It is more difficult to implant a bacterium in conventional animals than in axenic or antibiotic-treated animals. The barrier effect of the autochthonous microbiota against allochthonous and pathogenic species is known as colonization resistance. The competition for adhesion receptors, nutritional competition, and the production of inhibitory substances (antagonism) are among the mechanisms involved in reducing bacterial colonization and preventing bacterial overgrowth. Recently, an inverse correlation was found between the proportion of salivary bacteria inhibiting *Streptococcus mutans* and the percentage of untreated carious teeth, suggesting a possible role of inhibitory substances in the maintenance of oral health. Inhibitory substances include organic fatty acids, hydrogen peroxide, lactic acid, antibiotics, enzymes, and bacteriocins. The production of lactic acid by *S. mutans* and *Lactobacillus* generates low pH and inhibits the growth of *S. sanguis* and *S. oralis* as well as gram-negative bacteria. It has been suggested that the production of hydrogen peroxide by oral streptococci may reduce the growth of periodontopathogens. Bacteriocins are bactericidal proteinaceous substances, and those that are produced by streptococci may have a wide spectrum of activity against other gram-positive bacterial species. It has been argued that bacteriocins play a limited role in the ecology of the oral cavity because they are probably rapidly inactivated by the numerous proteases found in saliva. However, it has been found that some bacteriocins are not affected by human saliva. Many in vivo experiments with humans and animals indicate that the production of bacteriocins confers an ecological advantage. It is easier to implant strains of *S. mutans* that produce bacteriocins in the oral cavity than non-bacteriocin-producing strains.

External Factors

Diet

It is well documented that frequent consumption of a high-sucrose diet enhances the development of *S. mutans* and *Lactobacillus*. The fermentation of sucrose into lactate generates a low pH, favoring acidogenic and acidophilic bacteria. The substitution of sucrose with weakly fermentable sugar alcohols such as xylitol results in a reduction of *S. mutans* numbers and caries. Xylitol appears to selectively inhibit carbohydrate metabolism in *S. mutans*, which reduces acid production and thereby stabilizes the composition of the oral microbiota. Apart from the effect of sucrose, very few studies have addressed the effect of other dietary components on the oral microbiota. During our research on the effects of various dietary components on the oral microbiota of mice, we found that a high-starch diet favored an increase in the proportion of *Enterococcus faecalis* while a high-protein diet favored an increase in *Lactobacillus murinus*. Variations in the vitamin, lipid and mineral content of the diet had no direct effect on the oral microbiota of mice.

Oral Hygiene and Antimicrobial Agents

Oral hygiene is one of the most important factors in the maintenance of oral homeostasis and oral health. The mechanical removal of plaque by tooth brushing and flossing can almost completely prevent caries and periodontal diseases. The addition of antimicrobial agents to dentifrices, mouthwashes, and varnishes increases the effect of mechanical oral hygiene procedures. Antimicrobial agents may assist in protection by reducing bacteria adhesion to the tooth surface, by reducing the growth of microorganisms and plaque accumulation, by selectively inhibiting only those bacteria directly associated with oral diseases, or by inhibiting the expression of virulence determinants, such as acid production or protease activity. Fluoride is found in most toothpastes and mouth rinses and is well known for its anti-caries properties. The principal caries-preventive effect of fluoride is attributed to the formation of fluoroapatite and calcium fluoride, which lead to an increase in the resistance of enamel to demineralization. It can also inhibit bacterial growth by reducing the sugar transport, glycolytic activity, and acid tolerance of many gram-positive species. Fluoride can help stabilize the composition of the microflora by reducing the rate of acid production and the fall in pH during frequent carbohydrate intake. Other agents that have been formulated for commercial toothpastes and/or mouth rinses include chlorhexidine, quaternary ammonium compounds, plant extracts, metal ions, and phenolic compounds. These antimicrobial agents have been shown to reduce dental plaque formation, caries, and gingivitis.

Drugs and Diseases

Salivary gland hypofunction and xerostemia may result from the intake of xerogenic medication, irradiation treatments for head and neck cancer, and Sjögren's syndrome. Patients with xerostomia have a decrease capacity to eliminate sugar and buffer the acids found in plaque. In addition to suffering from the reduction in saliva protection, patients with xerostomia generally consume soft, high-sucrose diets and suck sour candies to keep their mouths moist. Patients suffering from xerostemia have higher levels of mutans streptococci, lactobacilli, staphylococci, and *Candida*, while the levels of *S. sanguis*, *Neisseria*, *Bacteroides*, and *Fusobacterium* are reduced compared with those in a healthy individual. They are also more susceptible to dental caries and candidiasis.

Antibiotics that are given orally or systemically for the treatment of different infections may enter the oral cavity via saliva and gingival crevicular fluid and lead to a imbalance in the oral microbiota. Antibiotics may suppress some resident bacterial populations which can result in overgrowth of antibiotic-resistant bacteria, infections by opportunist pathogens such as *Candida*, and colonization by exogenous potential pathogens such as yeasts and members of the *Enterobacteriaceae.*

Other Factors

Many other external factors may affect the oral microbiota; these include the wearing of dentures or partial dentures, smoking, oral contraceptives usage, malnutrition, host macroenvironment, and various exposures to exogenous bacterial species.

2

Bacteriology

BACTERIA

Bacteria are unicellular free living organisms without chlorophyll having both DNA and RNA. They are capable of performing all essential processes of life like growth, metabolism and reproduction.

The size of bacteria is so small that they are measured in microns. The shapes of bacteria are classified under:

1. *Cocci:*
   Cocci in cluster–staphylococci
   Cocci in chain–streptococci
   Cocci in pair–diplococci
   Cocci in groups of four–tetrad
   Cocci in groups of eight–sacrina
2. *Bacilli:* They are cylindrical or rod shaped
3. *Vibro:* These are comma shaped, curved rods and derived their name from characteristic vibratory motility.
4. *Spirochates:* They are relatively longer, thin, flexible having several coils.
5. *Actinomycetes:* They are branching filamentous bacteria resembling to radiate sunrays.
6. *Mycoplasma* are organisms which lack cell wall and do not possess a stable morphology.

Gram-positive organisms are not decolorized and they retain the colour of basic stain while gram negative lose their basic stain when treated with decoloriser and take up the counterstain.

INFECTIONS DUE TO GRAM-POSITIVE ORGANISMS

1. Scarlet fever
2. Diphtheria

3. Tuberculosis
4. Leprosy
5. Actinomycosis
6. Tetanus
7. Syphilis
8. Botryomycosis.

INFECTIONS DUE TO GRAM-NEGATIVE ORGANISMS

1. Tularemia
2. Melioidosis
3. Gonorrhea
4. Cat scratch disease
5. Noma

INFECTION DUE TO GRAM-POSITIVE ORGANISMS

Scarlet Fever

It is a systemic infection produced by group A β-haemolytic streptococci. The disease begins as a streptococcal tonsillitis with pharyngitis in which the organism elaborate an erythrogenic toxin that attacks the blood vessels and produces characteristic skin rash. Local erythematous reaction occurs in susceptible individuals injected with toxins. The incubation period is 3–5 days.

Clinical Features

Commonly seen in children from age 3 to 12 years. The microorganisms enter through the pharynx and exhibits severe pharyngitis, tonsillitis, headache, chills, fever and vomiting which is accompanied by enlargement and tenderness of regional cervical lymph nodes. The characteristic diffuse, bright red scarlet rash appears on second or third day of illness. The rash is prominently seen in areas of skin folds is called 'pasta lines'.

Oral Manifestations

Stomatitis scarlatina mucosa of palate appear congested and have scattered petechiae on soft palate. The palate appears fiery red with faucial pillars swollen with grayish exudates.

The first 2 days shows white coating through which only fungiform papillae can be seen this has been called white strawberry tongue.

The fourth or fifth day red strawberry tongue which is due to desquamation of white coat which reveal an erythematous dorsal surface with hyperplastic fungiform papillae which is also known as "red raspberry tongue".

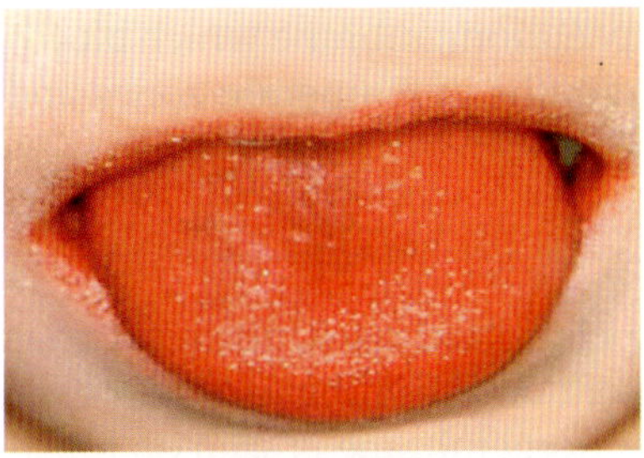

Strawberry tongue in scarlet fever

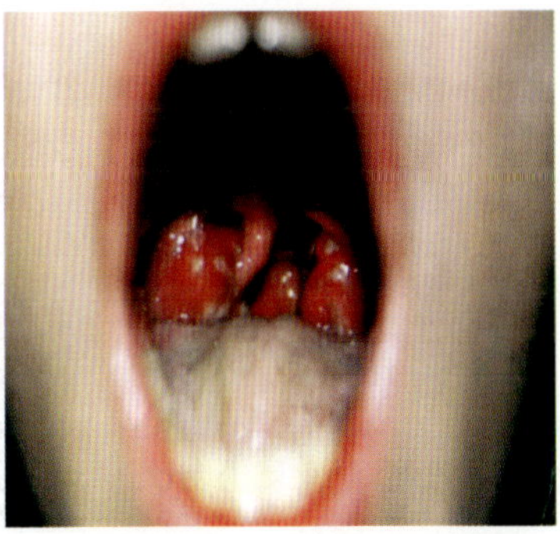

Tonsillitis and pharyngitis

Diagnosis

Culture of throat secretions to confirm streptococcal infection.

Treatment

Antibiotics like pencillin, dicloxacillin, cephalexin help in controlling possible complications.

Ibuprofen can be used to reduce fever and relieve associated discomfort.

Diptheria

It is a life-threatening acute infections and communicable disease of skin and mucous membrane caused by toxemic strains of *Corynebacterium diphtheriae.*

This is characterized by local inflammation and formation of greyish adherent pseudomembrane which bleeds on removal. This occurs in winter months in temperate zones and throughout year in tropical region.

Clinical Features

The patients are classified as diphtheria cases if the pseudomembrane is present and diphtheria carriers if pseudomembrane is absent. The pseudomembrane is seen on tonsils which is a wash leather elevated, greyish green membrane with a well-defined edge surrounded by acute inflammation. The incubation periods for respiratory diphtheria are 2–5 days and up to 8 days. Cutaneous diphtheria is a secondary infection on preexisting skin lesions.

There may be swelling of neck (bull neck) and tender enlargement of lymph nodes. The onset is gradual and manifests as fever, sorethroat, weakness, dysphagia, tachycardia.

Oral Manifestations

The formation of "diphtheritic membrane" which often begins on tonsils and enlarges becoming confluent over the surface. This membrane is greyish, green thick, fibrous gelatinous appearance.

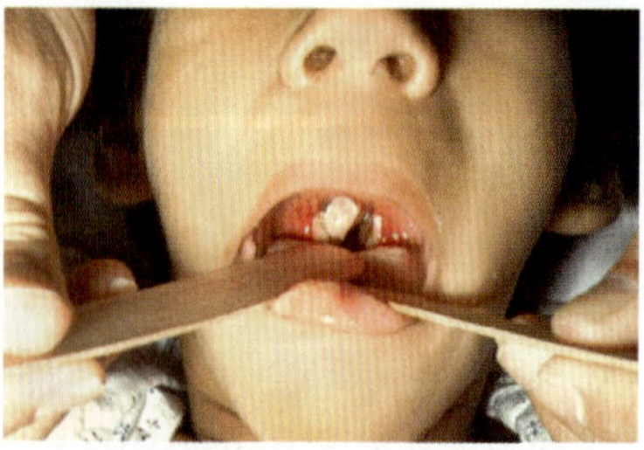

Tonsillitis

The submandibular and anterior cervical nodes are enlarged with soft tissue edema which gives bull-neck appearance.

Diagnosis

This is done mainly based on clinical signs and symptoms but definite diagnosis is done by isolation of organisms from the affected site.

Treatment

The disease may be prevented by prophylactic active immunization with diphtheria toxoid.

Once disease is developed it is treated with antitoxin usually in combination with antibiotics.

Tuberculosis

Tuberculosis is a specific infectious granulomatous disease caused by *Mycobacterium tuberculosis*. This affects lungs but also affects the intestine, meninges, bones, joints, lymph glands, skin and other tissues of body.

Etiology and Pathogenesis

This is a chronic infectious disease caused by *Mycobacterium tuberculosis*. Infection must be distinguished from active disease.

Primary Tuberculosis

Occurs in previously unexposed people and almost involves the lungs. Most infections are result of direct person to person spread through airborne droplets from a patient with active disease. The organism elects a non-specific chronic inflammatory reaction. In about 5–10 percent patients TB progresses from infection to active state and immunosuppression is often responsible. Active disease usually develops later in life from a reactivation of organisms which is called secondary tuberculosis.

Miliary Tuberculosis

Diffuse dissemination through vascular system may occur and often produces multiple small foci of infection that grossly and radiographically resemble millet seeds.

Clinical Features

Primary TB is usually asymptomatic. Secondary lesions are located at the apex of lungs but may spread to many different sites by expectorated infected material or through lymphatic or vascular channels. Extrapulmonary tuberculosis is seen and represents an increasing proportion. Any organ system may be involved including lymphatic system, skin, skeletal system, CNS, kidneys. Involvement of skin may develop and called as lupus vulgaris. Head and neck involvement may occur. The most common extrapulmonary sites in head and neck are cervical lymph nodes followed by larynx and middle ear.

Oral Manifestations

Painless ulcers less frequently nodular, granular or leukoplakic areas may be seen. The majority of lesions represent secondary infection from initial pulmonary foci.

This may involve gingiva, mucobuccal fold and areas of inflammation adjacent to teeth or in extraction sites, secondary oral lesions are mostly present on tongue, palate and lip.

Non-tuberculous mycobacterial infection from contaminated milk are rare but it is termed as scrofula.

Histopathology

Areas of infection demonstrate formation of granulomas, which are circumscribed collection of epithelioid histiocytes, lymphocytes and multinucleated giant cells often with central caseous necrosis. This is termed as tubercle.

Special stains such as Ziehl-Neelsen or other acid-fast stains are required to demonstrate the mycobacteria.

Diagnosis

Organisms are demonstrated by special microbial stains and culture of infected tissue or sputum. Presence of AFB in sputum smear is gold standard for diagnosis.

CT scan is used to diagnose mediastinal or hilar lymphadenopathy, cavities and intralesional calcifications.

Microscopy smears for AFB are stained by Z-N staining, Kinyoun's cold staining method and rhodamine staining for fluorescent microscopy.

- Tuberculin test

Treatment

Multiple drug therapy is often recommended INH along with rifampcin for 9 months or INH along with rifampcin and pyrizamide for 2 months followed by INH and rifampcin for 4 months.

Leprosy

Leprosy is a chronic infectious disease produced by *M.leprae*. This disease is slightly contagious. It mainly affects skin, peripheral nerves, upper respiratory tract, eyes and testes but also muscles, bones and joints.

Etiology

M.leprae is a obligate, intracellular gram-positive, acid-fast bacillus. It is only bacteria infecting peripheral nerve. It grows best in cooler tissues like skin, peripheral nerves, upper respiratory tract, anterior chamber of eyes and testes.

Pathogenesis

An immunologic and epidemiological study suggests most people develop a subclinical infection and very few develop infection. Once infected both cell mediated and humoral responses are elicted by bacterial antigen DNA glycolipids .

The bacteria are taken by histiocytes in skin and Schwann cells in the nerves. This usually results in an inflammatory response involving histiocytes and lymphocytes. It is clinically called indeterminate type.

Clinical Features

Bimodal age distribution with peaks at 10–14 and 35–44 years. This exists in 2 polar forms namely:

1. Tuberculoid type and
2. Lepramatous type.

General features are hypopigmented patches, partial or total loss of cutaneous sensation in affected areas. The thickening of nerves and presence of acid-fast bacilli in skin or nasal smear are common.

Tuberculoid lesions are characterized by single or multiple macular, erythematous eruptions with dermal nerve and peripheral nerve trunk involvement resulting in loss of sensation and loss of sweating on affected site.

Facial paralysis occur with some frequency due to nerve involvement.

Oral Manifestations

Small tumor like masses called lepromas which develop on tongue, lips or hard palate. These nodules show a tendency to break down and ulcerate.

Gingival hyperplasia with loosening of teeth are reported.

Diagnosis

Clinical and bacteriologic examination, nasal smears and scrapings stained with modified Ziehl-Neelsen method to detect leprae bacilli at a concentration greater than 10/11 gm of tissue.

Skin biopsy, nerve biopsy and foot culture histamine test.

Treatment

- Long-term chemotherapy is initiated upon diagnosis.
- Rifampcin and dapsone for six months in case of tuberculoid type.
- Rifampcin and dapsone along with clofazimine in case of lepromatous type.

Actinomycosis

It is a chronic granulomatous suppurative and fibrosing disease caused by anaerobic gram-positive non acid-fast branched filamentous bacteria.

The most commonly isolated organisms is *A.israeli*, *A.bovis* produces 'lumpy jaw'.

The usual disease is characterized by formation of abscesses that tend to drain by formation of sinus tracts. If pus from abscesses is examined on a clean glassslide it shows typical "sulfur granules" or colonies of organisms which appear in suppurative material.

Types

1. Cervicofacial
2. Abdominal
3. Pulmonary

Pathogenesis

Disruption of mucosal barrier is the main step in the invasion of bacteria. Initial acute inflammation is followed by chronic indolent phase. Lesions appear as single or multiple.

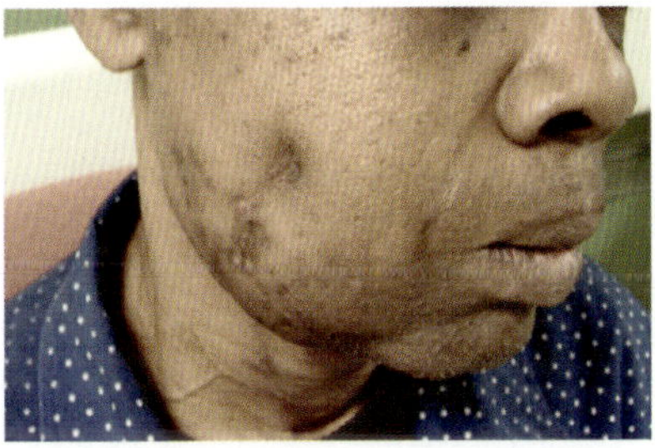

Actinomycosis infection in face

Cervicofacial

This is the most common form and the organisms enter the tissues through oral mucous membranes and remain localized in the subadjacent soft tissues or spread to involve the salivary glands, bone or even skin of face and the neck. The soft tissues swellings eventually develop into one or more abscesses and pus containing "sulfur granules" are seen. The overlying skin is purplish red, indurated and has feel of wood or often fluctuant. The infection may extend to involve maxilla or mandible. The destruction of tissues may be exclusive when bone is reached.

Abdominal

Serious form and carries high mortality rate. In addition fever, chills, accompanied by nausea and vomiting are seen. Involvement of liver and spleen are seen.

Pulmonary

Fever, chills accompanied by producing cough and pleural pain. They spread beyond lungs to involve adjacent structures.

Diagnosis

Diagnosis ideally achieved by culture. Lacking positive culture results, a strong presumptive diagnosis through demonstration of typical colonies in lesional biopsy material.

Treatment

Long-standing fibrosis cases are treated by draining abscesses, sinus tracts with high doses of antibiotics.

Pencillin and tetracycline have been used more frequently.

Tetanus (Lock Jaw)

Tetanus is an acute infection of nervous system characterized by intense activity of motor neurons resulting in severe muscle spasms. It is caused by exotoxin of anaerobic gram positive bacillus *C.tetani* which acts as a synapse of interneurons of inhibitory pathways and motor neurons to produce blockade of spinal inhibition.

Pathogenesis

1. In newborns where umbilical cord is cut with an unsterile instrument.
2. In children with otorrhea.
3. In women with unsterile handling of genital tracts.
4. Acute trauma.

Under suitable anaerobic conditions with low oxidation-reduction potential spores of *C. tetani* germinate and produce toxins. Once released toxins enter peripheral motor nerve terminal enters axons and is transported to nerve cell body in the brainstem and spinal cord by retrograde intraneuronal transport.

The toxins migrate across the synapse to presynaptic terminale. The length of the nerve traveled determines the time of ascent and explains why disease manifests first in muscles.

Clinical Features

This has an incubation period of 6–10 days.

Generalized Tetanus

This is characterized by lock-jaw or trismus due to spasm of masseter which is the initial symptom.

Dysphagia, stiffness or pain in the neck shoulder or back muscles appear concurrently. Laryngeal spasms leads to asphyxia. Rigid abdomen and stiff proximal limb muscles. Hands and feet are relatively spared and sustained contraction of facial muscles.

Local Tetanus

This manifests as spasms of muscles near wound. Cephalic tetanus is characterized by trismus and facial palsy is rare. The acute trismus which may develop in these patients similate acute oral infections, trauma, TMJ dysfunction and even hysteria.

Treatment

The aim of treatment is to remove the spores at the site of wound, prevent toxin production, neutralize unbound toxins and prevent muscular spasms. Cardiopulmonary monitoring should be maintained continuously. Sedation, airway obstruction should be maintained.

Antibiotics should be given to eradicate vegetative organisms or source of toxins.

Pencillin 10–12 million units IV for 10 days and metronidazole 1 gm every 12 hours should be administered.

Antitoxin is used to neutralize circulating toxin and unbound toxins within wound.

TIG 3000–6000 units IM individual doses, a dose of 500 unit is effective.

ATS–1500 units can be given.

Syphilis

This is caused by *Treponema pallidum*, a spirochete and characterized by episodes of active disease interrupted by period of latency.

Gram-positive, motile, microaerophilic spirochete and is characterized by episodes of active disease interrupted by period of latency demonstrated by dark field microscope since it stains poorly except silver impregnation.

Acquired Syphilis

Three distinct stages:

1. Primary stage
2. Secondary stage
3. Tertiary stage

Primary Stage

A lesion known as chancre develops at site of inoculation approx 3–90 days after contact with infection. This is usually solitary but may be multiple at times.

Seen commonly in penis of male; vulva or cervix of female. Of particular interest it occurs in lips, tongue, palate, gingival and tonsils. The usual primary lesion is an elevated, ulcerated nodule showing local induration and producing regional lymphadenitis. The intraoral chancre is an ulcerated lesion covered by a greyish white membrane which may be painful due to secondary infection. The chancre appears microscopically as a specific ulcer showing a rather inflammatory infiltrate. Plasma cells are numerous.

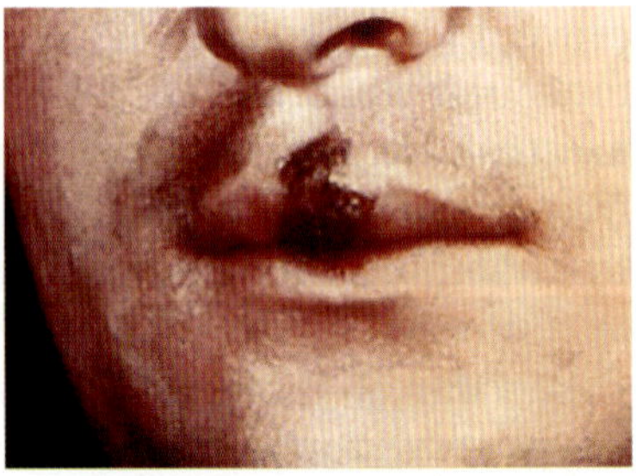

Chancre at primary stage

Secondary or Metastatic Stage

This usually commences after 6 weeks of primary lesion, characterized by diffuse eruptions of skin and mucous membranes. The lesions are multiple, painless, greyish white plaques overlying an ulcerated surface. They occur more frequently on tongue, gingiva or buccal mucosa.

Secondary syphilis can present as an explosive and wide spread form known as lues malinga characterized by fever, head ache and

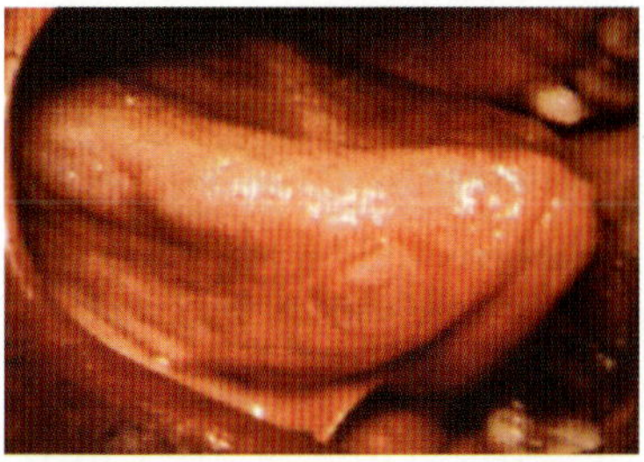

Affected tongue in secondary syphilis

muscle pain followed by necrotic ulcerations. After 2nd stage patients are free from lesions and symptoms they enter the latent stage which may last for 1–30 years.

Tertiary Syphilis

This is also called late syphilis and involve CVS, CNS and other tissues and organs. This is more infectious. The classical lesion is called gumma. This occurs more frequently in skin amd mucous membrane, liver, testes and bone. This consists of a focal, granulomatous inflammatory process with central necrosis. The initial oral gumma is seen on tongue and palate. The lesion appears as a firm, nodular mass in tissue which may frequently ulcerate to form a deep painless ulcer. Lesions of palate cause perforation by sloughing of necrotic tissues.

Congenital Syphilis

This is transmitted to the offspring only by an infected mother before their 4th month of pregnancy.

1/3–aborted or stillbirth

1/3–normal children

1/3–congenital syphilitic children

Short maxilla, frontal bossing, high palate, saddle nose, mulberry molars, irregular thickening of sternoclavicular portion of clavicle.

Hutchinson's Triad

- Hypoplasia of incisors and molar teeth
- Eighth nerve deafness
- Interstial keratitis

Diagnosis and Treatment

Examining the exudates of active lesion under a dark field microscope for spirochetes.

Specific and highly sensitive serological tests are fluorescence treponemal antibody absorption.

Penicillin is drug of choice.

Erythromycin or tetracycline is used in patients allergic to penicillin.

Botryomycosis

This is a granulomatous disease seen in horses caused by actinomycetes.

Clinical Features

Human botryomycosis is usually a localized granulomatous infection of skin or mucosa. This may disseminate involving liver, lungs or kidneys. The significance is that localized infection may mimic actinomycosis by producing granulomatous mass with multiple ulcers and sinuses.

Treatment

Non-specific treatment as caused by many microorganisms.

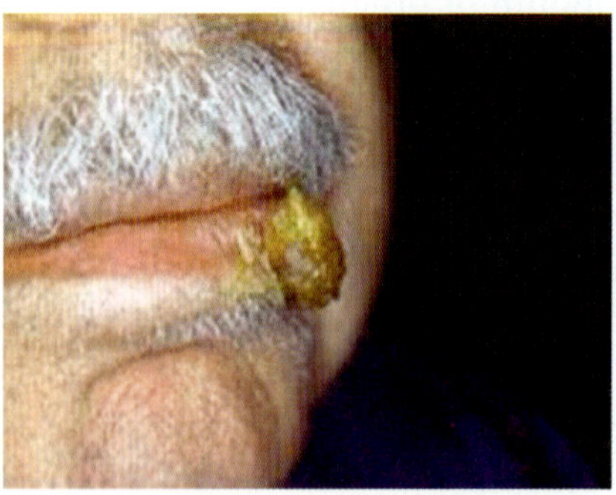

Botryomycosis in lower lip

GRAM-NEGATIVE ORGANISMS

Tularemia

This is caused by bacillus *Francisella tularensis*. This infection is contracted through infected rabbits and rodents. The exposure occurs and subsequent infection during skinning and dressing freshly killed infected animals and their ingestion. This is highly communicable from infected mammals to humans.

Clinical Features

Several types exist and they include cutaneous, ophthalmic, pleuropulmonary, oral and abdominal forms. After the incubation period of 7 days the patient suffers from sudden headache, nausea, vomiting, chills and fever. A cut or sore on skin develops into a suppurative ulcer. The lymphatic become enlarged, swollen and

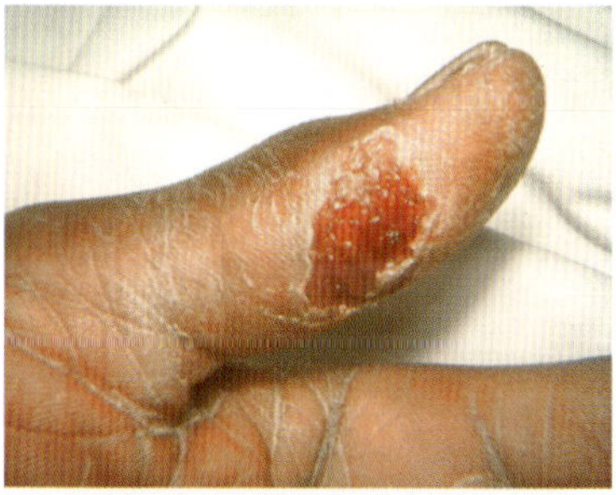

Tularemic infection in thumb

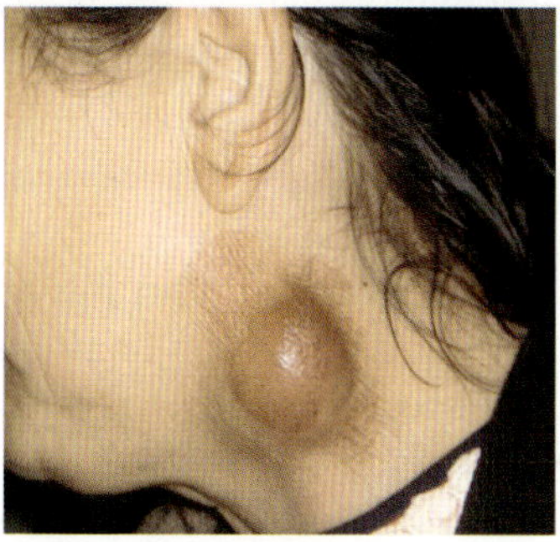

Tularemic infection on neck

painful. The eyes become involved with conjunctivitis developing through localization of the disease.

Tularemic pneumonia and pleuritis are complications which may eventuate in gangrene and lung abscess.

Oral Manifestations

Necrotic ulcers of oral mucosa or pharynx accompanied by severe pain. Some develop stomatitis single nodular masses eventually develop into abscesses. Regional lymphadenitis are seen in submaxillary and cervical group of nodes.

Treatment

Streptomycin is the drug of choice. Tetracycline may be used.

Melioidosis

Specific infection in men and animals caused by bacillus *Pseudomonas pseudomallei* an aerobic gram –ve, non-acid fast and rod-shaped bacilli. The disease is endemic.

Clinical Features

Two recognized form of disease exist

1. Acute form
2. Chronic form

Acute Form

The patients develop high fever, evidence of acute pulmonary infection, diarrhea and hemoptysis. There is a wide spread visceral involvement as a result of hematogenous dissemination. Death occurs due to septicemia.

Chronic Form

The chronic form is seen in those who have survived the acute form of disease. Granulomatous type characterized by multiple, small, non-specific abscesses occurring subcutaneously or in viscera, lymph nodes or bones.

The mode of infection occurs through contamination of soil or water.

Treatment

Incision and drainage accompanied by tetracycline or in combination with chloramphenicol.

Gonorrhea

This is a veneral disease affecting male and female genitourinary tract and is transmitted by sexual intercourse. It is an infection of epithelium and manifests as cervicitis, urethitis, prostatitis and conjunctivitis. This is caused by *Neisseria gonorrhoeae* a gram –ve, non-motile, non-spore forming organisms that grows in pairs.

The common age is 15–25 years. This has morbidity in developing countries and plays a role in HIV transmission. This has a incubation period is 1–5 days.

Clinical Features

Male

It results in acute urethritis, dysuria and urethral discharge of purulent in nature. It may lead to epididymitis, chronic prostates, balantitis and posterior urethritis.

Female

Cervicitis with candidal or trichomonal vaginitis. Mostly asymptomatic. This may affect rectum, oropharynx and eyes.

Oral Manifestations

This is not only confined to homosexuals and occurs as a result of oro genital contact or inoculation through infected hands. The lips

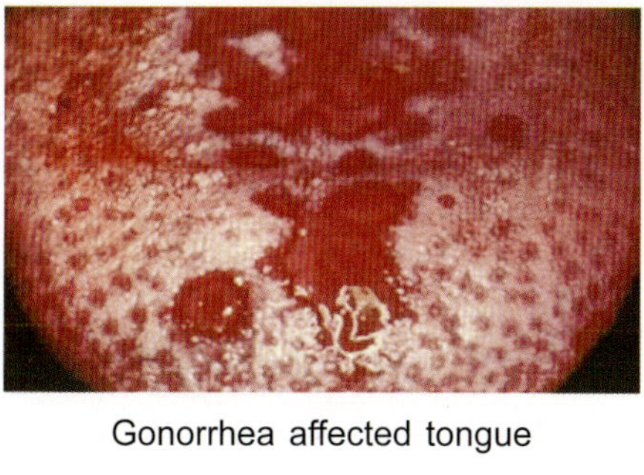

Gonorrhea affected tongue

may develop acute painful ulcerations limiting motion, gingival may become erythematous, lips develop painful ulcerations. Oral lesions are accompanied by fever and regional lymphadenopathy.

Complications

Epididymitis, pelvic inflammatory disease.

Diagnosis

Gram staining of urethral discharge samples can be collected with Dacron or rayon swabs.

Treatment

Antibiotics and preventive measures.

Cat Scratch Disease

The causative organisms is *Bartonella henselae*. Gram –ve stained with silver stain.

Clinical Features

The disease occurs at any age but predominant in children and young adult due to traumatic break in skin due to scratch or bite of household cat.

The primary lesion is often a papule, pustule or vesicle that develops at site of injury. Within 1 or 3 weeks a regional lymphadenitis without lymphangitis develop. The cat serves as a carrier. The nodes are painful and may be several centimeters in diameter with inflammation of overlying skin. This persists for 1 to 6 weeks accompanied in early stage by low-grade fever, chills, malaise or even abdominal pain.

If preauricular, submaxillary or cervical chain of nodes are involved the dentist may be consulted to rule out dental disease as an etiology.

Diagnosis

Positive intradermal skin test of antigen from a patient which proved cat scratch disease.

Indirect immunofluorescent antibody assay is used to detect antibodies for *Bartonella*.

Prognosis

Its good as disease is self limiting and regresses within a week or month. Incision and drainage of involved nodes. Antibiotic therapy is ineffective.

Noma

Noma, which means to devour is a rapidly spreading mutilating, gangrenous stomatitis that occurs usually in debilitated or nutritionally deficient patient it is seen chiefly in children, but also found in adults under certain conditions, such as those existing among the malnourished individuals.

Predisposing Factors

1. Persons who are under nourished or debilitated from infections such as diphtheria, dysentery, measles, pneumonia, scarlet fever, syphilis, tuberculosis and blood dyscrasias, including anemia.
2. Noma may be considered a secondary complication of systemic disease rather than a primary disease.
3. Noma appears to originate as a specific infection by vincent's organisms, and acute necrotizing gingivostomatitis, which is soon complicated by secondary invasion of many other microbial forms, including streptococci, staphylococci and diphtheria bacilli.

Clinical Features

Noma usually begin as a small ulcer of the gingival mucosa which rapidly spreads and involves the surrounding tissue of the jaws, lips and cheeks by gangrenous necrosis.

The Initial site is commonly an area of stagnation around the fixed bridge or crown.

The overlying skin becomes inflamed, edematous and finally necrotic, with a result that a line of demarcation develops between healthy and dead tissue, and large masses of the tissue may slough out leaving the jaw exposed.

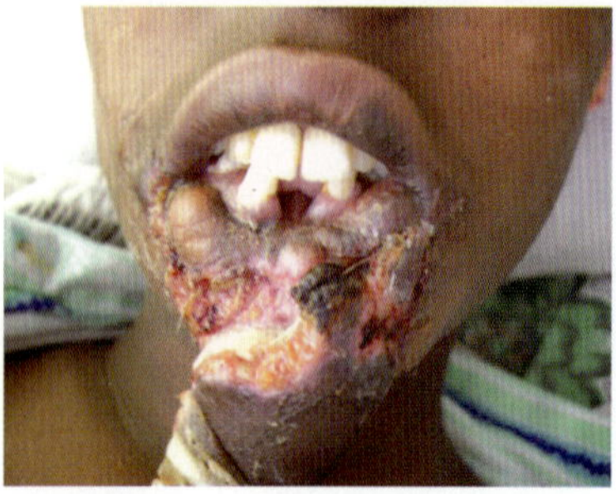

Noma

The order arising from the gangrenous tissue is extremely foul. The patient have high temperature during the course of the disease, suffer secondary infection and may die from toxemia or pneumonia.

Treatment

Antibiotics: Immediate treatment of any existing malnutrition further improves the probability of saving the patient.

3

Dental Caries

INTRODUCTION

Dental caries is an irreversible microbial disease of the calcified tissues of the teeth, characterized by demineralization of inorganic portion and destruction of organic substance of the tooth which often leads to cavitation. It affects persons of both genders in all races, all socio-economic strata and every age group.

Etiology

Role of Carbohydrates

The presence of readily fermentable carbohydrates has been thought to be responsible for the loss of caries resistance. A cariogenicity of a dietary carbohydrate varies with the frequency of ingestion, physical form, chemical composition, route of administration and presence of other food constituents. Sticky, solid carbohydrates are more caries producing than consumed as liquids. Polysaccharides are less easily fermented by plaque bacteria than monosaccharides and disaccharides. Plaque organisms produce little acid from the sugar alcohols, sorbitol and mannitol. Glucose or sucrose fed entirely by stomach tube or intravenously does not contribute to any decay as they are unavailable for microbial breakdown. Refined, pure carbohydrates are more caries producing than crude carbohydrates complexed with other food elements capable of reducing enamel solubility or possessing antibacteria properties.

Role of Microorganisms

Miller demonstrated the presence of microorganisms within the tubules of decayed teeth, mainly cocci and leptothrix laid the foundation for the role of acids elaborated by bacteria in caries production. It played a role in decalcification of both enamel and dentin. From the carious dentin isolated bacteria called bacillus acidophilus odontolyticus was found. A new *Streptococcus mutans* was invariably isolated from carious lesions in the teeth. Microorganisms isolated from the deeper carious cavities were mainly acidogenic streptococci and concluded that there was an apparent relationship of lactobacilli with initial caries and of streptococci with more advanced lesions of dentin.

Microbial Flora and Dental Caries

Type of caries	*Microorganisms*
Pit and fissure	*S. mutans*
	S. sanguis
	Lactobacillus sp
Smooth surface	*S. mutans*
	S. salivarius
Root surface	*A. viscosus*
	A. naeslundii
	S. mutans
	S. sanguis
Deep dentinal caries	*Lactobacilli* sp
	A. naeslundii
	Filamentous rods

Role of Acids

Mechanism of carbohydrate degradation to form acids in oral cavity by bacterial action probably occurs through enzymatic breakdown of sugar, acids formed chiefly lactic acid. Carbohydrates and sugar have implicated the caries etiology and acid production from these food substances. Anaerobic catabolism of carbohydrates called fermentation or glycolysis predominates in plaque. The end products of glycolysis have the same empirical formula as starting substrate in that more molecule of glucose breaks into two molecules of lactic acid. Organisms such as streptococci and many lactobacilli ferment sugars which produce 90% or more lactic acid as end product are called homofermentative. Heterofermentatives produce a mixture of metabolites including organic acids such as propionic, butyric, succinic and ethanol.

Tooth Factor

The structure and composition of teeth undoubtedly influence the initiation and rate of progression of a carious which determines the caries susceptibility or caries immunity. Studies on chemical composition of enamel indicate that surface enamel is more resistant to caries than subsurface enamel. Surface enamel is more highly mineralized and tend to accumulate greater quantities of fluoride, zinc, iron than underlying enamel. On the other hand, the concentration of carbonate, magnesium and sodium is lower in the surface layer and increases near DEJ.

Theories

The proteolysis-chelation theory

Theory

Simultaneous microbial degradation of the organic components and the dissolution of the minerals of the tooth by the process known as chelation.

Chelation

A process involving the complexing of a metallic ion to a substance through a covalent bond which results in a highly stable, poorly dissociated or weekly ionized compound.

Effects of Chelation

Independent of pH of the medium

That removal of metallic ions such as calcium from a biologic calcium phosphorus system may occur at a neutral or even alkaline pH.

The proteolysis-chelation theory resolves the argument as to whether the initial attack of dental caries is on the organic or inorganic portion of enamel by stating that both may be attacked simultaneously.

Composition of the Tooth

The structure and composition of teeth undoubtedly influence the initiation and the rate of a progression of a carious lesion, which determines its caries susceptibility or caries immunity.

The Proteolytic Theory

Theory: Organic or protein elements of tooth as initial pathway of invasion by microorganisms.

Enamel lamellae as pathways for organisms in the progress of dental caries.

Gottlieb and Gottlieh, Diamond and Applebaum:

Caries is essentially a proteolytic process: The microorganisms invade the organic pathways and destroy them in their advance. Acid formation accompanied proteolysis.

Drawbacks and Conclusion

No satisfactory evidence to support the claim that the initial attack on enamel is proteolytic.

Gnotobiotics studies: Caries can occur in the absence of proteolytic organisms.

Conclusion

Proteolysis in the initiation of dental caries is likely to be of no significance but its role in the progression of the more advanced carious lesions cannot be ruled out.

Summary of Miller's Experiment

An essential role to three factors
- The oral microorganisms
- The carbohydrate substrate
- The acid

Questions Unanswered by Miller

- Predilection of specific sites on a tooth
- The initiation of smooth surface caries
- Why some populations are caries free
- The phenomenon of arrested caries.

Mechanisms of Bacterial Adherence

Bacteria appears to posses surface components that have recognition potential and which bind to specific receptors on the pellicle and other host tissues. These surface components are referred to as adhesions. Some adhesions bind to saccharide receptors. Protein adhesions, which bind to specific sugars, are called 'lectins'. Other adhesions which contain hydrophobic moieties, may interact with hydrophobic residues receptor. Adhesions, therefore, permit bacterial cells to recognize and adhere to complex macromolecules.

Most of carbohydrates in the matrix consists of polymers glucans, fructans and heterosaccharides synthesized by the bacteria.

The high concentration of calcium and phosphate in plaques is thought to be due, in part to the infiltration of salivary proteins containing these constituents into the bound form. These probably include staherin, the salivary protein which, by adsorbing onto early crystal nuclei and preventing crystal growth, maintains super saturation of the fluid phase of plaque with apatite.

It is now known that the formation of dental plaque requires two types of specific bacterial adherent interactions, firstly bacteria attach selectively to acquired pellicle and secondly, bacteria accumulate via specific adhesive and cohesive interactions involving components of the plaque matrix and direct bacterial cell contact.

The pellicle appears as three distinct structures. The subsurface component, below of enamel having anaerobic catabolism of carbohydrate called fermentation of glycolysis predominates in plaque. The end products of glycolysis have the same empirical formulas as the starting substrate in that one molecule of glucose breaks into two molecules of lactic acid.

Organisms such as streptococci and many lactobacilli ferment sugars which produce 90% or more lactic acid on the end product such bacteria are called homofermentative.

Heterofermentatives produce a mixture of metabolities including into organic acids such as propionic, butyric, succine and ethanol.

Role of the Dental Plaque

Dental plaque, microbial plaque or bacterial plaque is a structure of vital significance which was demonstrated for the first time in histological preparations by Williams in 1987. It is believed to consist of such salivary components as mucian desquamated epithelial cells and microorganisms. It characteristically forms on tooth surfaces which are not constantly cleansed and appears as a tenacious thin film which may accumulate to a perceptible degree in 24–28 hours. An important component of dental plaque is acquired pellicle, which forms just prior to or concomitantly with bacterial colonization and may facilitate plaque formation. The pellicle is a glycoprotein that is derived from the saliva and is adsorbed on tooth surfaces. It is not dependent on bacteria but may serve as a nutrient for plaque microorganisms.

The pleasure of plaque, however, does not necessarily mean that a carious lesion will develop at that point variations in caries formation have been attributed to the nature of the plaque itself, to the saliva or to the tooth. Ability to adhere to smooth surfaces.

Plaque is the soft, non mineralized bacterial deposit which forms on teeth and dental prostheses that are not adequately cleaned. A characteristic of plaque is that it resists removal by physiologic and oral cleansing forces such a saliva and tongue movement but is removable by tooth brushing. Evidence that bacteria are the major component of plaque is derived from studies showing approximately 2 × 10 bacteria per gram. Plaque is composed of about 80% water and 20% solids. Bacterial and salivary proteins comprise about one half of the dry weight of plaque.

Role of Acids

The mechanism of carbohydrate degradation to form acids in the oral cavity by bacterial action. Probably occurs through enzymatic breakdown of the sugar, and the acids formed are chiefly lactic acid. Although others such as butyric acid to do form.

Microorganisms isolated from the deeper carious cavities were mainly acidogenic streptococci and thus concluded that there was an apparent relationship of lactobacilli with initial caries and of streptococci with more advanced lesions of dentin.

Miller's Chemoparasitic Theory or Acidogenic Theory

Theory

Caries is caused by acids produced by microorganisms of mouth

Dental decay is a chemoparasitic process consisting of two stages.

- Decalcification of enamel and dentin (preliminary stage)
- Dissolution of the softened residue (subsequent stage)

Acids resulting in primary decalcification are produced by the fermentation of starches and sugar from the retaining centres of teeth.

Role of Carbohydrates

The presence of readily fermentable carbohydrates has been thought to be responsible for their loss of caries resistance.

The carcinogenicity of dietary carbohydrate varies with the frequency of ingestion, physical form, and chemical composition, route of administration and presence of other food constituents. Sticky, solid carbohydrates are more caries producing than those consumed as liquids.

Polysaccharides are less easily fermented by plaque bacteria than monosaccharides and disaccharides. Plaque organisms produce little acid from the sugar alcohol, sorbitol and mannitol. Glucose or sucrose, fed entirely by stomach tube or mannitol. Glucose or sucrose, fed entirely by stomach tube or intravenously does not contribute to decay as they are unavailable for microbial break down.

Refined, pure carbohydrates are more caries producing than crude carbohydrates complex with other food elements capable of reducing enamel solubility or possessing antibacterial properties.

Role of Microorganisms

Miller demonstrated the presence of microorganisms within the tubules of decayed teeth mainly cocci and leptothrix, as he called them and laid the foundation for the role of acids elaborated by bacteria in caries production.

Played a role in decalcification of both enamel and dentin. From carious dentin they isolated bacteria which they called *Bacillus acidophilus* odontolyticus. A new *Streptococcus* species, *S. mutans*, which was almost invariably isolated from carious lesions in the teeth of British patients.

Several studies have shown that by the age of six years, about 20% of children have experienced dental caries in the their dentition and a DMFT of 0.5 can be expected, by age of 12 years 90% of children would have experienced a DMFT accelerates at a greater rate than the DMFT beyond the age of eight years. So that by age 12, a DMFS of 7.5 can be taken as an average figure.

Another common element is that children from families in lower socioeconomic element is that children from greater in lower prevalence than their peers from families at a higher socioeconomic level.

Permanent teeth are greater in females than in males of the same age. This is attributable largely to the fact that the teeth of girls erupt at an earlier age. The caries experience in deciduous teeth is greater in males. This time difference is particularly significant during the formative years because teeth have been shown to be maximally susceptible to dental caries immediately after eruption. This is due to the fact that the chemical structure of teeth in the immediate post eruptive stage is suboptimal in terms of caries resistance.

Siblings of individuals with high caries susceptibility are also generally caries active, whereas siblings of caries immune individuals generally exhibit low caries rates. Children of parents with a low caries experience also tend to have low caries; the converse is true for children whose parents have a high caries rate.

Types of Caries

Clinical Classification of Caries

According to the anatomical site

- Pit or fissure caries
- Smooth surface caries

According to the rate of caries progression

- Acute dental caries
- Chronic dental caries

According to the nature of attack

- Primary (virgin) caries
- Secondary (recurrent) caries
- Based on chronology
- Infancy caries (rampant caries)
- Adolescent caries.

Deciduous dentition. It has been variously attributed to prolonged use of:

- A nursing bottle containing milk formula, fruit juice or sweetened water
- Breastfeeding
- Sugar or honey–sweetened pacifiers.

The carious process in affected teeth may be so severe that only root stumps remains.

Adolescent Caries

There are two chronological periods when acute, rapidly progressing caries is commonly observed. Acute exacerbations in caries rates are usually seen at 4–8 years of age and at 11–18 years of age. The acute caries attack in the latter period is usually characterized as adolescent caries. There are usually seen in teeth and surface that are relatively immune to caries. With a relatively small opening in enamel with extensive undermining of enamel. Because of rapid progression of lesion there is little time for the formation of reparative dentin. It is important to detect cases of rampant caries in the adolescent at an early stage so that prevention procedures may be rigorously applied.

Chronic Dental Caries

Chronic dental caries is that form which progresses slowly and tends to involve the pulp much later that acute caries. It is most common in adults. The entrance to the lesion is almost invariably larger. Because of this there is not only less food retention, but also greater access of salvia. The slow progress of the lesion allows sufficient time for both sclerosis of the dentinal tubules and deposition or reparative dentin in response to adverse irritation. The carious dentin is often stained deep brown.

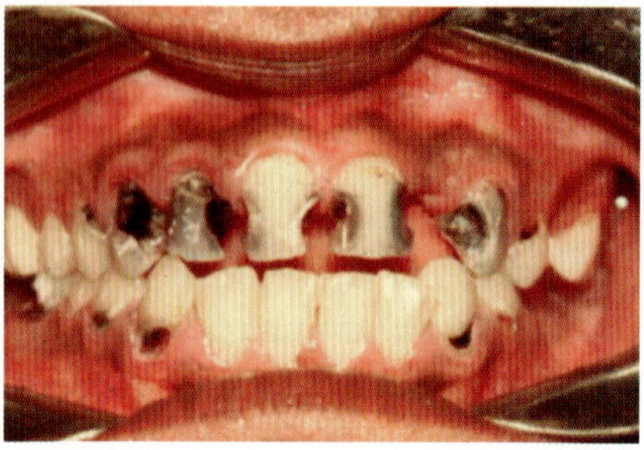

Chronic caries lesion in anterior teeth

Recurrent Caries

Recurrent caries is that type which occurs in the immediate vicinity of a restoration. It is usually due to inadequate extension of the original restoration, which favors rentention of debris, or to poor adaptation of the filling material to the cavity, which produces a leaky margin. In either event, the renewed caries follows the same general pattern as primary caries.

Arrested Caries

Arrested caries has been described as caries which becomes static and does not show any tendency for further progression. The deciduous and permanent dentitions are both affected by this condition. It occurs almost exclusively in caries of occlusal surfaces and is characterized by a large open cavity in which there is lack of food retention and in which the superficially softened and decalcified dentin is gradually burnished until it takes on a brown stained, polished appearance and is hard. This has been referred to as 'eburnation of dentin'. Sclerosis of dentinal tubules and secondary dentin formation commonly occur in cases of arrested caries.

Typical Acquired Characteristics

- The presence of brown pigmentation
- The change from a soft to a hard texture

Classification based on Severity and Rate of Progression

Acute Dental Caries

Acute dental caries is that form of caries which runs a rapid clinical course and results in early pulp involment by the carious process.

Rampant Caries

A condition which is characterized by sudden, rapid and almost uncontrollable destruction of teeth, affecting surface of teeth that we relatively caries free. These include the proximal and cervical surface of the mandibular incisors which are relatively caries free. A caries increment of 10 or more new carious lesions over a period of about a year is characteristic of rampant caries. Rampant caries are most often observed in the primary dentition of young children and the permanent dentition of teenagers. Dietary factors affecting oral substrate and oral flora and physiological factors affecting saliva are often significant in the development of rampant caries.

Nursing Bottle Caries

Also called nursing caries, baby bottle syndrome, and bottle mount syndrome, is an interesting and unfortunate form of rampant caries.

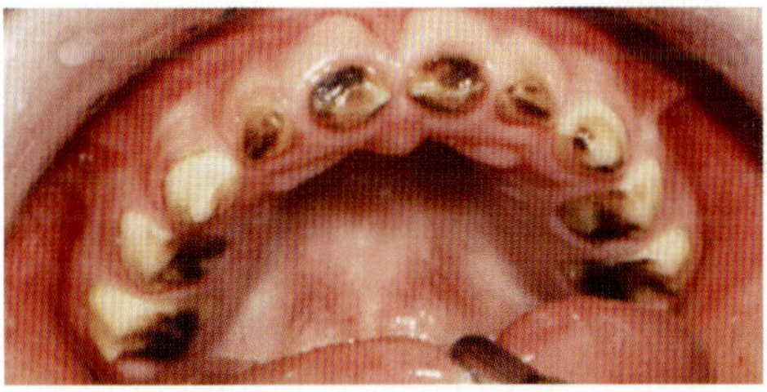

Nursing bottle caries

Root Caries

Root caries is defined by Hazen and his colleagues as "a soft progressive lesion that is found anywhere on the root surface that has lost connective tissue attachment and is exposed to the oral environment" root caries.

Root caries initiates on mineralized cementum and dentin surfaces which have greater organic component that enamel tissue. Root surface caries in contemporary populations occurs most frequently on the buccal and lingual surfaces of roots.

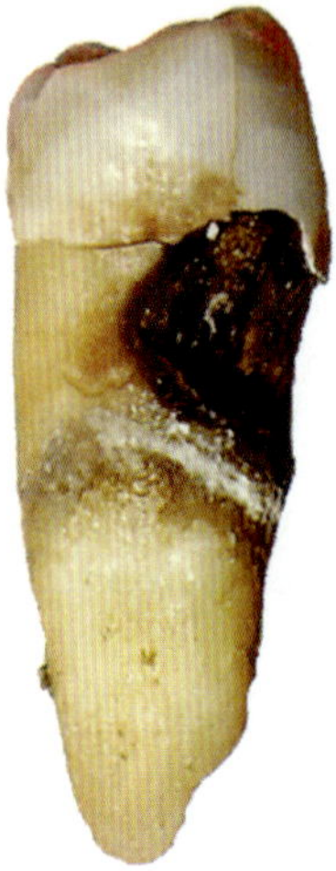

Root caries

Microorganisms appear to invade the cementum either along Sharpey's fiber or between bundles of fibre, in a manner comparable to invasion along dentinal tubules. Since cementum is formed in concentric layers and presents a lamellated appearance the microorganisms tend to spread laterally between the various layers. Irregular mineralization on this cemental surface may often be seen at the same time, probably representating beginning of calculus formation. After decalfication of the cementum, destruction of the remaining matrix occurs similar to the process in dentin, with ultimate softening and destruction of this tissue. As the caries process continues, there is invasion of microorganisms into underlying dentinal tubules, subsequent matrix destruction and finally pulpal involvement.

Steep walls and narrow bases are those most prone to develop caries due to their mechanical characteristics which result in poor self cleansing features. Sometimes consider development faults. Deep narrow pits and fissures favour the retention of food debris along with microorganisms and caries may result from fermentation of this food and the formation of acid.

Pits and fissures affected by early caries may appear brown of black and will feel slightly soft and catch a fine explorer point. The enamel directly bordering the pit or fissure may appear opaque bluish white as it becomes undermined. This undermining occurs through lateral spread of the caries at the detinoenamel junction, and it may be rapid process if the enamel in the base of the pit or fissure is thin.

The lateral spread of caries at the junction as well as penetration into the dentin the dentinal tubules may be extensive without fracturing away the overhanging enamel. Thus, there may be large carious lesion with only a tiny point of opening. This undermined enamel may suddenly give way under the stress of mastication, or the dentist may suddenly open into a large cavity when excavating the pit or fissure.

Smooth Surface Caries

Smooth surface caries of the primary type is caries that develops on the proximal surface of the teeth or on the gingival third of the buccal and lingual surfaces. Generally preceded by the formation of a microbial plaque. This ensures the retention of carbohydrate and microorganisms on the tooth surface in an area not habitually cleansed and the subsequent formation of acid to initiate the caries process.

Proximal caries usually begins just below the contact point, and appears in the early stage as a faint white opacity of the enamel without apparent loss of continuity of the enamel surface. The early white chalky spot becomes slightly roughened, owing to superficial decalcification of the enamel. As the caries penetrates the enamel, the enamel surrounding the lesion assumes a bluish-white appearance similar to that seen something around carious pits of fissures.

Linear Enamel Caries

An atypical form of dental caries that has been observed in the primary dentition of children in Latin Amercian and Asian countries. The lesions predominate on the labial surface of the anterior maxillary teeth, in the region of neonatal line, which results from metabolic disturbances such as hypocalcemia or trauma of birth. A variant of the linear enamel form of caries in the primary teeth of children in the Far East has been named adontoclasia. The morphological aspects of this type of caries are atypical and results in gross destruction of the labial surfaces of incisor teeth.

Cervical Caries

This type of caries occurs on buccal, lingual or labial surfaces and usually extends from the area apposing the gingival crest occlusally to the convexity of the tooth surface marking the self–cleansing portion of this surface.

Cervical caries is almost always an open cavity and does not present the narrow point of penetration seen commonly in pit or fissure caries and proximal caries.

Diagnosis of Dental Caries

Radiographic Diagnosis

Fatty degeneration of tomes fibres

- Terminology–questioned
- Disposition of fat globules–precedes early sclerotic changes
- Special stains–sudan red
- Significance
- Fat contributes to impermeability
- Predisposing factor for dental sclerosis.

Sclerotic dentin

- Reaction of vital pulp–calcification of DT
- Seals off DT from further penetration of microorganisims
- Minimal in rapidly advancing caries
- Prominent in slow caries
- Sclerotic dentin–appear white in transmitted light.

Decalcification of dentinal tubules

- Above dentinal sclerosis–zone of decalcification
- Occurs in advance of bacterial invasion of DT
- Pioneer bacteria
- The initial decalcification–only the walls of DT
- Study of tubules–pure form of microorganisms.

Zone of microbial invasion

- Proteolytic organisms–predominant in deeper layers
  Acidogenic organisms–more in early caries
- Supporting the hypothesis that intimation and progression are two distinct processes and must be differentiated.

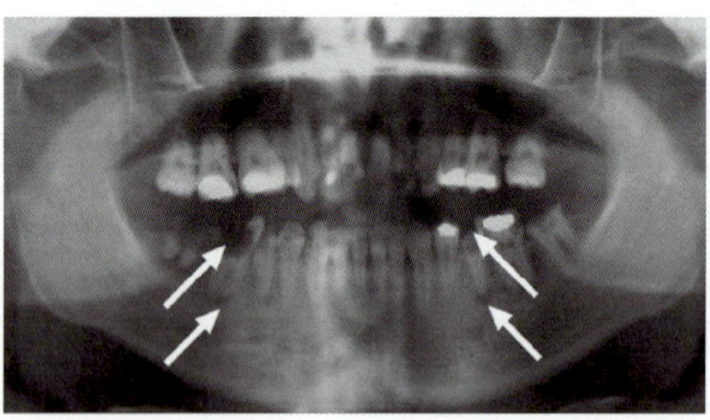

OPG showing multiple caries lesion. Infrared laser fluorescence, digital imaging fiberoptic, transilumination, early dentinal caries

Zones of Enamel Caries

Translucent Zone (TZ)

- First recognizable zone of alteration
- Advancing front of the lesion
- Half the lesions demonstrate TZ, not always present
- Seen in long GS in clearing (quinolone–RI–1.62)
- TZ appears structure less
- Pore volume–1% (compared to 0.1% of sound enamel).

Dark zone

- Lies adjacent and superficial to the translucent zone
- Positive zone
- Shows positive birefringence (in constract to sound enamel)
- Pore volume of 2–4% (polarized light)
- Presence of small pores–large molecules of quinolone are unable to prenetrate
- Micropore system–gets filed with air and become dark
- Medium like water may penetrate.

Body of the lesion

- Between unaffected surface and dark zone
- Area of greatest demineralization
- Pore volume–5% in periphery and 25% in centre
- Quinolone imbibition–body appears transparent
- Water imbibition ±ve birefringence compared to sound enamel
- Striae of Retzius–prominent.

Surface Zone

- Quantitative studies–partial demineralization of 1–10%
- Pore volume–less than 5% of the spaces
- Negative birefringence–water imbibition
- Positive birefringence–porous subsurface
- The change from a chalky whiteness to light brown.
- No increase in the size of the lesion
- No further progress of the lesion as long as the pigmentation remained.

Radiation Caries

The development of rampant caries in patients undergoing radiation therapy in the head and neck region is referred to as radiation caries. Xerostomia was a complication of radiation therapy and postulated that caries development may be due to a modification of saliva secretion. Besides xerostomia, an increase in viscosity and low pH saliva after irradiation were also observed.

As a consequence of irradiation, three forms of dental defects are described as explained by Frank, et al. 1965 and Baden, 1970. The first type is a characteristic caries like lesion usually completely encircling the neck of the tooth. The second type of lesion begins as brown to black discoloration of the crown. The occlusal surface of posterior teeth and incisal edges of anterior teeth wear away. The third type of lesion begin as a spot depression. Which spreads from incisal or occlusal edges on the labial or buccal and lingual surfaces.

Mechanical Measures for Caries Control

- Oral prophylaxis by the dentist
- Toothbrushing
- Mouthrinsing
- Use of dental floss or toothpicks
- Incorporation of detergent food in the diet.
- Plaque control agents

Antiseptic

1. Positively charged organic molecules
   a. Quaternary ammonium compounds acetylpyridinium chloride
   b. *Pyrimidines:* Hexidine
   c. Bis-bisquanides, chlorhexidine
2. Noncharged phenolic agents, listerine, triclosan, phenol and thymol
3. *Oxygenating agents:* Peroxides and perborates
4. *Bispyridinies:* Octenidine
5. *Halogens:* Iodine iodophors and fluorine
6. *Heavy metal salts:* Sliver, mercury, ainc, copper and tin

Antibiotics

Niddamycin, kanamycin sulfate, tetracycline hydrochloride, vancomycin hydrochloride.

Enzymes

- Mycinases, pancreatin, funal enzymes and proteases
- Plaque modifying agents
- Urea peroxide
- Sugar substitutes
- Xylitol and mannitol
- Plaque attachment interference agents
- Sodium polyvinylphosphoric acid and perfluoroalkyl.

Nutritional Measure for Caries Control

Microbiological examination provides an objective method for assessment of the patient cooperation. Changes in dietary habits are reflected within a couple of weeks in corresponding reductions in the number of oral lactobacilli and *Streptococcus mutans*.

The control of dental caries through nutritional or dietary means is possible to achieve on the basis of a mass prevention program and, for this reason, is relatively unimportant in public health and preventive dentistry in contract to fluoridation of water supplies.

The chief nutritional measure advocated for the control of dental caries is restriction of refined carbohydrate intake. Only the most cooperative patient will adhere rigidly to the type of diet designed to reduce sugar consumption drastically.

- Penicillin
- Erythromycin
- Kanamycin
- Spiramycin
- Tetracycline
- Tyrothricin
- Vancomycin
- Urea and ammonium compounds
- Chlorophyll
- Nitrofurans
- Sliver nitrate
- Zinc chloride and potassium ferrocyanide
- Substances which interfere with carbohydrate degradation through enzmatic alterations.

- Vitamin K
- Sarcoside
- Substances which interfere with bacterial growth and metabolism
- Fluoride mouthwashes or rinses
- Bis-Biguanides
- Chlorhexide and alexidine
- Sustances which interfere with bacterial growth and metabolism
- Fluoride dentrifices
- Topical application of fluoride
- Sodium fluoride
- Stannous fluoride (SnF_2)
- Fluoride supplements
- Quantitative light fluorescence.

Chemical Measures of Caries Control

A vast number of chemical substances have been proposed for the purpose of controlling dental caries. The use of some of these has been upon sound experimental evidence. Use of others has been purely empirical and without scientific foundation. These chemicals include:

- Substances which alter the tooth surface or tooth structure
- Substance which interface with carbohydrate degradation through enzymatic alterations
- Substances which interfere with bacterial growth and metabolism

In the light of our present knowledge, each of these may theoretically be of benefit in controlling caries. The final proof, however, depends upon through clinical trial.

Fluorine

Fluorine as a member of halogen family because of its electronegative properties makes it extremely reactive. In mineralized tissues such as bones and teeth, it occurs as the apatite salt of fluoridated hydroxyapatite. Fluorine has been administered principally in two ways: Though the communal water supply and by topical application. Fluoridation of water supplies.

4

Gingival and Periodontal Diseases

Acute Gingival Infections

Classification of varoius acute gingival lesions

According to Manson

a. Traumatic lesions of gingival:
   - Physical injury
   - Chemical injury

b. Viral infections:
   - Acute herpetic gingivostomatitis
   - Herpangina
   - Hand, foot and mouth diseases
   - Measles
   - Herpes varicella/zoster virus infections
   - Glandular fever

c. Bacterial infections:
   - Acute necrotizing ulcerative gingivitis
   - Tuberculosis
   - Syphilis

d. Fungal diseases
   - Candidiasis

e. Gingival abscess

f. Apthous ulceration

g. Erythema multiforme

h. Drug allergy and contact hypersensitivity.

Terminology

Acute necrotizing ulcerative gingivitis (ANUG) is an inflammatory, destructive disease of the gingiva, which presents characteristic signs and symptoms. Other terms used to describe this condition are:

- Vincent's infection
- Trench mouth
- Acute ulceromembranous gingivitis and others.

NUG (necrotizing ulcerative gingivitis) can cause destruction of the supporting structures. When it involves the bone it cause bone loss, the condition is referred as necrotizing gingivitis that can occur in acute, subacute and recurrent forms.

Clinical Features

Necrotizing ulcerative gingivitis is characterized by sudden onset sometimes may be followed by an episode of debilitating diseases or acute respiratory.

Tract infections long hours of working without adequate rest and psychologic stress are also frequent features in the history of necrotizing ulcerative gingivitis.

Intraoral Signs and Symptoms

a. Oral signs; lesions are characterized by punched out crater like depressions at the crest of the interdental papillae subsequently involving marginal gingiva and rarely attached gingiva.
b. These craters are covered by grayish pseudomembranous slough, which is demarcated from the remaining of the mucosa by a pronounced linear erythema.
c. These ulcerations of necrotizing ulcerative gingivitis could be of two types: lateral ulceration and necrosis deep ulceration and necrosis. This is related to the fact that the gingival, lateral margins and lateral parts of the papillae, while intracellular vessels supplying to the col and central portions of the papillae lateral ulceration, which is (less common) involving primarily the buccal wall of the papillae, margins and possibly the attached gingiva, occurs in the distribution of the lateral blood supply. Deep ulceration involves necrosis of the tissues of the embrasure

giving rise to typical truncated papillae, occurs in the distribution to intra-alveolar vessels (more common).

d. Other signs include gingival hemorrhage or pronounced bleeding on the slightest stimulation.

e. Fetid odor and increased salivation.

Oral Symptoms

1. The lesions are extremely sensitive to touch
2. Complains of a constant radiating, gnawing pain that is intensified by eating spicy or hot foods and chewing
3. There is a metallic foul taste, and the patient is conscious of an excessive amount of "pasty saliva".

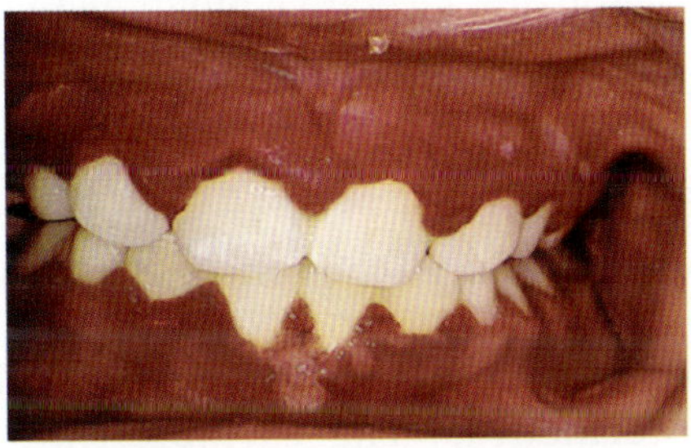

Severe gingivitis in ANUG

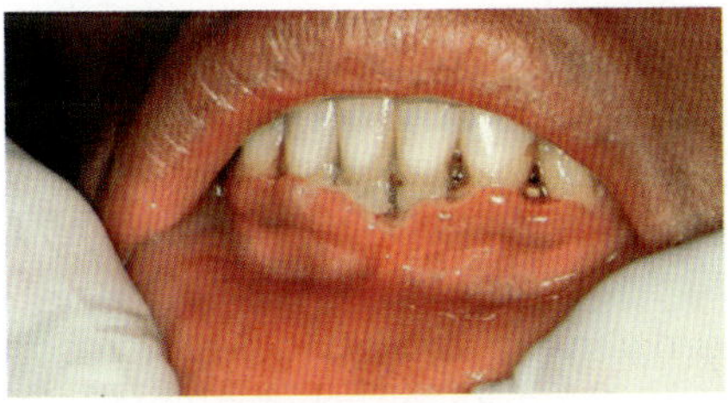

Gingivitis in relation to lower anteriors

Extraoral Signs and Symptoms

- In mild to moderate stages of the disease local lymphadenopathy and a slight elevation in temperature are common features.

- In severe case marked systemic are complications such as high fever, increased pulse rate, leukocytes loss of appetite and general lassitude is common systemic reactions could be more severe in children.
- In rare case, severe sequelae such as the following may occur noma or gangrenous stomatits, fusospirochetal meningitis peritonitis toxemia, and fatal brain abscess.

Clinical Course

It is indefinite, if left untreated it may lead to destruction of the periodontium and denudation of roots (necrotizing ulcerative periodontitis), combined with severe in toxic systemic complications.

Horning and Cohen have described the following stages in the progression of necrotizing ulcerative gingivits (NUG).

Etiology

Role of Bacteria

Plant and Vincent introduced the concept that acute necrotizing ulcerative gingivitis is caused by a specific bacteria namely a fusiform bacillus and a spirochetal organism and also bactericides intermedius. The variable flora consists of a heterogeneous array of fusospirochetal organism more recently loesche and colleagues described a content and a variable flora associated with ANUG.

Constant flora is composed of fusospirochetal organism and also bactericides intermedius and the variable flora consists of a heterogeneous array of bacterial types where bacteriologic findings have been supported by immunologic data increased lgG and lgM antibody titers to Spirochetes (intermediate sized up to 90%) and prevotella has been demonstrated. Electromicroscopic studies have been demonstrated, electron spirochetes, small intermediate sized (maximum number into 90%) and large spirochetes. The specific cause of acute necrotizing ulcerative gingivitis has not been established. The common opinion is that it is produced by a complex of bacterial organisms but requires underlying tissue changes. To facilitate the pathogenic activity of the bacteria.

Local Predisposing Factors

Most important predisposing factors are:

1. Pre-existing gingivitis
2. Injury to the gingiva
3. Smoking

Deep periodontal pockets, per coronal flaps are particularly vulnerable areas for the occurrence of the disease because they offer a favourable environment for the proliferation of the fusospirochetes and hence they are called as incubation zones.

Areas of the gingival traumatized by opposing teeth in malocclusion, such as the palatal surface behind the maxillary incisors and the labial gingival surface of the mandibular incisors are frequent sites of acute necrotizing ulcerative gingivitis.

Many investigators have reported a positive correlation between smoking and acute necrotizing ulcerative gingivitis. The possible reasons are:

- Direct toxic effect of tobacco on the gingiva.
- Vascular or other changes induced by nicotine or other substances.
- Smoking and acute necrotizing ulcerative gingivitis is both reflections of stress.

Systemic Predisposing Factors

Nutritional deficiency: A poor diet has been reported as a predisposing factor in ANUG. Nutritional deficiencies such as vitamin B and vitamin C accentuate deficiencies such as vitamin B and vitamin C accentuate the response of gingival tissues produced by increased pathogenic flora. Several researchers have found an increase in the fusospirochetal flora in patients with nutritionally-deficient diets.

Debilitating diseases: May predispose to the development of acute necrotizing ulcerative gingivitis. Such systemic disturbances are metallic intoxication, severe gastrointestinal disorders, blood dyscrasias such as anemia, leukemia and acquired immuno-deficiency syndrome.

Psychosomatic factors: Appears to be important in the etiology of acute necrotizing ulcerative gingivitis. The mechanisms whereby psychological factors create or predispose to gingival damage have not been established, but an alteration in digital and gingival capillary responses suggestive of increased autonomous nervous activity has been demonstrated. Cohen and coworkers have suggested that a psychiatric disturbance may lead to an activation of the hypothalamic pituitary adrenal axis. This results in elevation of serum and urine cortisol levels, which is associated with a depression of lymphocyte and polymorphonuclear leukocyte function that may predispose to acute necrotizing ulcerative gingivitis.

Histopathology

Microscopically, the lesions involve both epithelium and underlying connective tissue. The surface epithelium is destroyed and is replaced by a pseudomembranous mesh work of fibrin, necrotic epithelial cells, polymorphonuclear neutrophils and various types of microorganisms. This is the zone that appears clinically as the surface pseudomembrane. The underlying connective tissue is markedly hyperemic, with numerous engorged capillaries and dense infiltration of polymorphonuclear neutrophils. This acutely-inflamed hyperemic zone appears clinically as the surface pseudomembrane.

Aggresive Periodontitis

Clinical Features

Age and Sex Distribution

Affects both the sexes and is seen mostly between puberty and 20 years of age. Some studies show predilection to female patients.

Distribution of Lesions

Three areas of localization of bone loss have been described:

1. First molar and/or incisors.
2. First molar and/or incisors + additional teeth (not exceeding 14 teeth).
3. Generalized involvement.

For localized juvenile periodontitis classic distribution is in the first molars and incisors with least destruction in the cuspid, premolar area.

Limitations of destruction to certain teeth could be for the following reasons:

1. Production of opsonizing antibodies against *A. actinomycetemcomitans* called "burn-out" phenomenon.
2. Bacteria antagonistic to *A. actinomycetemcomitans* may develop thereby decreasing the number of colonization sites.
3. *A. actinomycetemcomitans* may loose its leukotoxin producing ability for unknown reasons.
4. Localization of the lesions could also be due to the defect in cementum formation (hypoplastic/aplastic cementum).

Clinical Findings

1. The most striking feature is lack of clinical inflammation despite the presence of deep periodontal pockets.
2. There is a small amount of plaque, which forms a thin film on the tooth and rarely mineralizes to become calculus.
3. Most common initial symptoms are mobility and migration of first molars and incisors. Classically, a distolabial migration of the maxillary incisors with diastema formation occurs, lower incisors rarely migrate compared to upper incisors, all changes followed by sequelae of migration are seen.
4. As the disease progresses other symptoms like root surface sensitivity, deep dull radiating pain, periodontal abscess formation and regional lymph node enlargement may occur.

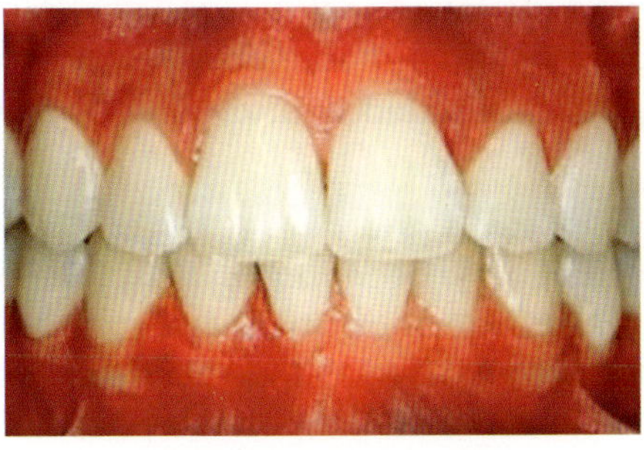

Aggressive periodontitis

Radiographic Findings

1. Vertical or angular bone loss around the first molars and insicors in an otherwise healthy teenagers is a diagnostic sign of classical juvenile periodontitis. The pattern appears to be,"Arc-shaped loss of alveolar bone extending from distal surface of 2nd premolar to mesial surface of 2nd molar".
2. Frequently,bilaterally symmetrically patterns of bone loss occurs, called as "mirror image pattern".

Pathogenesis of aggressive periodontitis is due to an interplay of several factors these include, the specific microbiology of subgingival plaque, defects in cementum, hereditary factors, impaired PMNs function and disorders of the immune system.

Histopathology/Microscopic Features

These are the same as those seen during pocket formation.

1. Like ulcerated pocket epithelium.
2. Accumulation of various inflammatory cells in the connective tissue mainly leukocytes, plasma cells and small number of lymphocytes and macrophages.
3. Electron microscopic studies of juvenile periodontitis revealed bacterial invasion of connective tissue that reaches the bone surface.
4. The flora involves *A. actinomycetemcomitans*, *Capnocytophaga sputigena* and others.

Bacteriology

Two types of bacteria are considered to be pathogens in localized aggressive periodontitis: *A. actinomycetemcomitans* is a short, facultatively anaerobic, non-motile gram-negative rod.

Virulence factors associated with *A. actinomycetemcomitans* are:

Immunology

Immune defects that have been implicated in the pathogenesis of localized aggressive periodontitis are functional defects of polymorphonuclear leukocytes/monocytes thereby it impairs the chemotactic attraction of PMNLs(polymorphonuclear leukocytes) to the site of infection.

Treatment

Prognosis is no more considered as poor patients with aggressive periodontitis. The following treatment has been tried in the past with varying results:

1. *Extraction:* Extraction of involved teeth especially first molar results in unevently healing.

Chronic Periodontitis

Introduction

Chronic periodontitis was previously known as "adult periodontitis" or "slowly progressive periodontitis".

Definition

Chronic periodontitis occurs as a result of extension of inflammation from the gingiva into the deeper periodontal tissues. It has recently been defined as "an infectious disease resulting in inflammation within the supporting tissues of the teeth, progressive attachment loss and bone loss".

Diagnostic Criteria for Chronic Periodontitis

Clinical features

1. Age of onset is usually 30 to 35 years.
2. The disease is usually generalized, although some areas are more deeply involved than the other areas.
3. No consistent pattern of distribution of lesion is seen, except that they are usually not isolated to one or two sites.
4. Highly acute inflammatory sites are not seen, mostly gingiva appear to be slight to moderately swollen and color may range from pale-red to magenta.
5. Loss of stippling, blunt or rolled gingival margins and flattened or cratered papillae may be seen.
6. Spontaneous bleeding and inflammation related exudate from the pockets may also be found.
7. When the pocket occludes it may result in abscess formation.
8. Pocket depths are variable, and both suprabony and infrabony pockets can be found.

9. Conditions that enhance plaque accumulation like open interdental contacts, defective restorative margins and malposed teeth may be frequently seen.
10. The amount of microbial deposits are consistent with severity of the disease.
11. Tooth mobility is seen in advance cases.
12. No serum neutrophil/monocyte abnormalities are seen.

Microbiological Features

Causative organisms of chronic periodontitis are:

1. *Porphyromons gingivalis* (*P. gingivals*)
2. *Prevotella intermedia* (*P. intermedia*)
3. *Capnocytophagia*
4. *A. actinomycetem comitans* (*A.a*)
5. *Eikenella corrodens* (*E.corrodens*)
6. *Campylobacter rectus* (*C.rectus*)

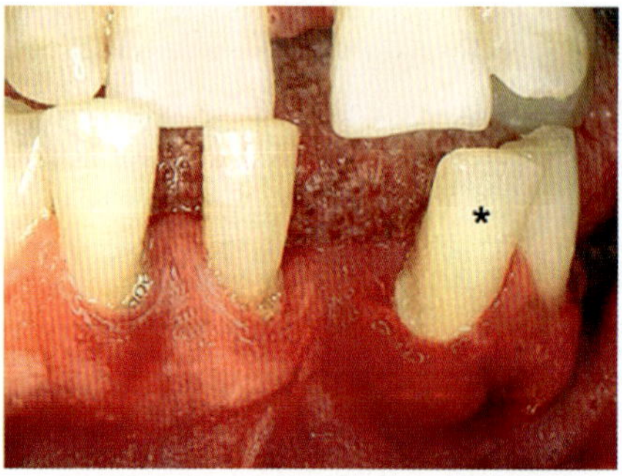

Aggressive periodontal condition

Radiographic Features

Pattern of bone loss observed in chronic periodontitis may be vertical or horizontal. When attachment loss and bone loss on one tooth surface is greater than that on an adjacent surface is referred to as vertical bone loss and is usually associated with angular bony defects and infrabony pockets. When attachment loss and bone loss occur at a uniform rate on majority of tooth surfaces it is called horizontal bone loss and is generally associated with suprabony pockets.

Types

Types are based on the following:

Disease Distribution

Localized periodontitis: Periodontitis is considered localized when less than 30 percent of the sites assessed in the oral cavity demonstrate attachment loss and bone loss.

Generalized periodontitis: It is considered generalized when more than 30 percent of the sites assessed in the oral cavity demonstrate attachment loss and bone loss.

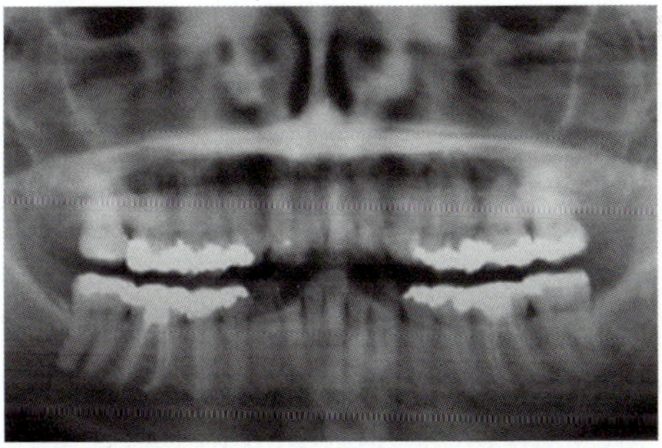

OPG showing bone loss in periodontal conditions

Disease Severity

Slight (mild) periodontitis:

1. Periodontal destruction is generally considered slight when there is not more than 1 to 2 mm of clinical loss of attachment.
2. Usually generalized involvement.
3. Minimal furcation invasions with or no mobility are observed.
4. Bone loss is minimal (less than 20% of total attachment).

Moderate Periodontitis

It is considered moderate when there is:

1.3 to 4 mm of clinical attachment loss.

NECROTIZING ULCERATIVE PERIODONTITIS, REFRACTORY PERIODONTITIS AND PERIODONTITIS AS A MANIFESTATION OF SYSTEMIC DISEASE

Necrotizing Ulcerative Periodontitis

Necrotizing ulcerative periodontitis occurs as a result of extension of necrotizing ulcerative gingivitis into the periodontal structures, leading to loss of attachment and bone loss. There are two types of necrotizing ulcerative periodontitis described, based on its relationship to acquired immunodeficiency syndrome(AIDS): Non-AIDS type necrotizing ulcerative periodontitis and AIDS-associated necrotizing ulcerative periodontitis.

Non-aids Type Necrotizing Ulcerative Periodontitis

Clinical Features

Since necrotizing ulcerative periodontitis occurs after repeated attacks of necrotizing ulcerative gingivitis (NUG) all the characteristic clinical features of necrotizing ulcerative gingivitis are seen, i.e.

- Ulceration and necrosis of gingival margin, which gets covered by a pseudomembranous slough.
- The ulcerated margins are surrounded by an erythematous halo.
- The lesions are extremely painful and bleed spontaneously.
- Localized lymphadenopathy, fever and malaise. These lesions, especially in long-standing cases, can extend to the deeper periondontal structures resulting in deep, crater-like osseous lesions especially in interdental areas. Such cases are identified as necrotizing ulcerative periodontitis (NUP), most striking feature of this condition is absence of deep conventional pockets associated with deep interdental osseous craters. This is because the necrotizing and ulcerative properties of the gingival lesion destroys the marginal epithelium, resulting in total destruction of the marginal tissue leading to recession.

AIDS-associated Ulcerative Periodontitis

Gingival and periodontal lesions of HIV positive patients appear to have similar findings that are seen in non-AIDS-associated necrotizing ulcerative periodontitis patients, in addition, they may exhibit certain complications such as:

- Large areas of soft tissue necrosis with exposure of bone and sequestration of bone.
- Sometimes these lesions may extend onto the buccal vestibule or the palate and become necrotizing stomatitis.

Refractory Periodontitis

According to American academy of periodontology, refractory periodontitis has been defined as "those cases which do not respond to any treatment provided, whatever the thoroughness or frequency". It should be differentiated from the cases of recurrent periodontitis, in which, after remission of the disease, recurrence follows due to inadequate plaque control either by the patient or clinician. Hence recurrent periodontitis, which is recurrence of the disease due to incomplete treatment should be differentiated from refractory cases whose reasons for not responding to adequate treatment, in certain cases, is still not clearly understood.

ETIOLOGY

Risk Factors Responsible for Refractory Cases

- Abnormal host response.
- Resistant strains of pathogenic periodontal microflora.
- Failure to eliminate plaque retentive factors, such as furcation involvement, irregular root surface, palatogingival groove, etc. which may in turn interface with complete plaque removal.
- On the other hand, smoking and systemic diseases may result in generalized lesions, which may not respond favourably to treatment.

Specific microorganisms have been identified in lesions of refractory periodontitis. Haffajee, *et al* (1988) reported three major microbial complexes in refractory periodontitis cases.

- *B. forsythus, F. nucleatum* and *C. rectus.*
- *S. intermedius, P. gingivalis* and *P. micros.*
- *S. intermedius* and *F. nucleatum.*

Clinical Features

In the classification proposed by the American academy of periodontology (1999), the term refractory periodontitis has been removed as a single entity due to the diversity of clinical conditions and treatments under which periodontal therapy fails to arrest the progression of periodontitis. Therefore the participants of the workshop have concluded that, rather than a single disease entity, the 'refractory' designation could be applied to all forms of periodontitis, e.g. refractory chronic periodontitis, refractory aggressive periodontitis and others.

Hence, refractory periodontitis in no way differs from other forms of periodontitis. Magnusson, et al (1991) noticed no changes in the amount of plague in relation to the sites that have been treated, however the sites that showed attachment loss had also showed persistent bleeding on probing. The refractory causes can be identified either by presenting new areas of attachment loss or progressive attachment loss in already treated sites.

Treatment

Antimicrobial therapy along with mechanical debridement has been found to be effecting in reducing the microbial population. Antibiotic therapy aims to reinforce mechanical periodontal therapy and also host defense system by killing subgingival pathogens that remain after conventional mechanical periodontal therapy.

Drugs used are

Amoxicillin, tetracycline hydrochloride, clavulanate potassium.

Periodontitis as a Manifestation of Systemic Disease

Severe periodontitis has been observed in patience who exhibits either defective numbers of neutrophils or defective neutrophils function. Some conditions are:

- Papillon-Lefevre syndrome
- Chediak-Higashi syndrome
- Down syndrome
- Hypophosphatasia

- Neutropenia
- Leukocyte adhesion deficiency.

Papillon-Lefevre Syndrome

It is characterized by hyperkeratotic skin lesion and severe destruction of the periodontium

These changes may appear before the age of 4 years. Skin lesion are hypokeratosis of localized areas on palms, soles, knees and elbows.

Periodontal involvement is early inflammatory changes that lead to bone loss and exfoliation of teeth.

Primary teeth are lost by 5 or 6 years of age

The permanent definition erupts normally but within few years the permanent teeth are also lost.

Chediak-Higashi Syndrome

This is a rare syndrome characterized by recurrent bacterial infections it exhibits oral ulceration and rapid destructive periodontitis.

Chediak-Higashi syndrome is hereditary disease with defect in both neutrophils and monocytes.

Down Syndrome

This is a congenital disease caused by a chromosomal abnormality and is characterized by mental disease and growth retardation. The oral findings include presence of plaque, calculus and other local factors like diastema, crowding of teeth, high frenal attachment and malocclusion.

Periodontitis in Down syndrome include the formation of deep periodontal pocket associated with plaque accumulation and moderate gingivitis, usually generalized but more severe in the lower anterior region; marked recession is also seen in this region.

Hypophosphatasia

This is a rare familial skeletal disease characterized by rickets, poor cranial formation, premature loss of primary dentition particularly incisors. Patient has low level of serum alkaline phosphatase. Teeth are lost with no clinical evidence of gingival inflammation and show reduced cementum formation.

Neutropenia

It is a condition where the circulating neutrophils are reduced that is less than 1500/ml destructive generalized periodontal lesions have been destroyed in children with neutropenia. Neutropenia could be inherent or acquired. Certain drugs and infections can lead to reduction in number of neutrophils.

Leukocyte Adhesion Deficiency

These cases are rare and begin during immediately after eruption of the primary teeth. Extreme acute inflammation and proliferation of gingival tissue accompanied by rapid bone loss is found, profound defects in peripheral blood neutrophils and monocytes are seen. Hence they are absent in gingival tissues. Patients with LAD also have frequent respiratory tract infection otitis media.

5

Virology

HERPES SIMPLEX

Acute infectious disease most commonly affecting man. There are immunologically two types of HSV infections. They are HSV1 and HSV2. A state of latency and reactivation is common in many viral infections, especially few herpes group. The incubation period is 1–26 days and can occur throughout the year. Transmission is mainly through close contact, kissing, sharing of glasses.

Pathogenesis: Mucosal surfaces and abraded skin favour entry and initiation of replication in the cells of epidermis and dermis. Initial or primary infection is asymptomatic and occurs commonly in childhood or infancy. Once the virus gains its entry it enters into sensory, autonomic nerve endings and remain latent in ganglia. Virus replication occurs and spreads to other skin and mucosal surfaces by centrifugal distribution of virions through peripheral nerves. This way of spread is responsible for wide and large surface involvement. The disease manifests as characteristic vesicles containing desquamated cells, multinucleated giant cells and free viruses with edema fluid. Though both subtypes produce orofacial and genital lesions which are indistinguishable. HSV1 mainly affects lips, face, oral cavity and upper body skin. HSV 2 usually affects the genitalis and skin of lower half of body. Primary infection resolves and the virus can no longer be recovered from ganglia but viral DNA will be seen in ganglion cells.

Principally two types of infections are seen in herpes, the first is a primary infection in a person who do not have the circulating antibodies and the second is a recurrent infection in persons having those antibodies.

HERPES GENITALIS

This is caused by HSV2 relatively common disease of uterine cervix, vagina, vulva and penis. This virus differs immunologically from type 1 which is responsible for oral infections. Type 2 is virulent compared to type 1.

Primary Herpetic Stomatitis

The virus is isolated from patients suffering from gingivostomatitis.

Clinical Features

Herpetic stomatitis is a common oral disease transmitted by droplet spread or contact with lesion. It affects both children and young adults. Rarely seen before 6 months of age due to circulating antibodies. Primary attack seen in children by development of fever, irritability, headache, pain upon swallowing and regional lymphadenopathy. Within few days the lips become painful and gingiva which is intensely inflamed appear erythematous and edematous. The lip, tongue, buccal mucosa, palate, pharynx and tonsils may be involved. Yellowish fluid vesicles develop which rupture and form shallow ragged extremely painful ulcers covered by a grey membrane and surrounded by a erythematous halo. Gingival inflammation proceeds formation of ulcers. Ulcers heal spontaneously leaving no scar. HSV1 can be isolated from facial, labial and oral herpetic lesions for a mean duration of three and half days with a range of 2–6 days, while HSV2 can be isolated from genital lesions for a mean duration of five and half days with a range of 2–14 days.

Herpes Infection in Lips

Mode of Transmission

Droplet infection, direct contact. After such contact incubation period ranges from 2 to 20 days with an average of 6 days before development of lesions.

Diagnosis

Scrapings obtained from base of lesions are stained with Wright's and Giemsa stain. PAP demonstrates balloon cells, multinucleated

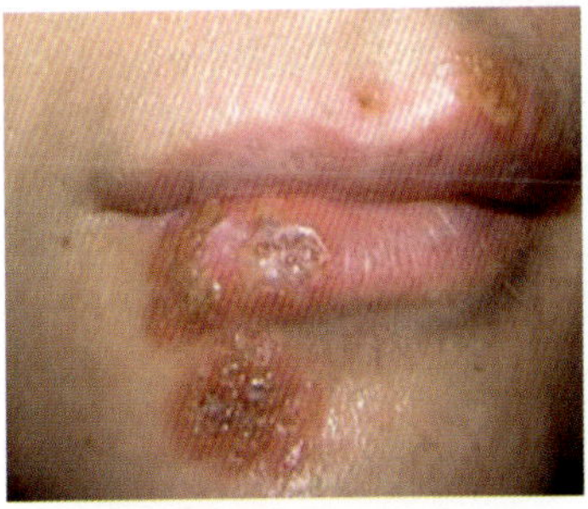

Herpes infection in lips

giant cells and intranuclear inclusions. Most accurate and sensitive method is PCR technique.

Treatment

Antiviral have good results, antibiotics helps to prevent secondary infection, NSAIDs and anesthetic gels relieve discomfort.

Recurrent or Secondary Herpetic Labialis and Stomatitis

This is usually seen in adult patients and manifests clinically as attenuated form of primary disease. The virus once entered appear to reside dormantly within the regional ganglia and when reactivation is triggered spread along nerves to sites of oral mucosa and skin where they destroy the epithelial cells and induce typical inflammation.

Clinical Features

Infection occurs in wide varying intervals from nearly every month to about once in a year or even less in others. The lesions may develop either at site of primary inoculation or in adjacent area supplied by involved ganglion. It may develop on lips, intraorally causing tingling sensation and tautness, swelling or slight soreness at location in which vesicles subsequently develop. These vesicles are small, tend to occur in clusters and may coalesce to form large lesions. These rupture quickly leaving red ulcers with a erythematous halo. The lesions heal in 7–10 days leaving no scar.

Multiple Ulcers in Palate

Lab Diagnosis

Isolation of virus can be done in tissue culture.

Immunofluorescent staining, immune peroxidase technique, serologic assays such as complement fixation, radioimmunoassay and ELISA.

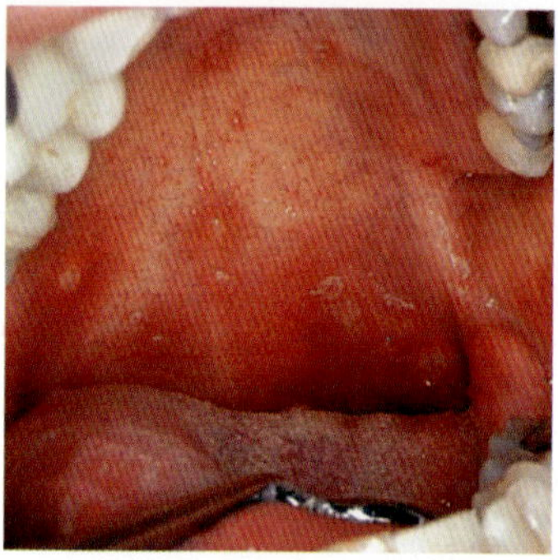

Multiple ulcers in palate

Treatment

Antiviral drugs are useful in treating.

Foot and Mouth Disease (Apthous fever, foot and mouth disease, epizootic stomatitis)

Foot and mouth disease is a viral infection, which only rarely affects man, but does affect dogs and sheep as well as cattle. Transmission of diseases occurs through contact with infected animals; in human beings this is usually through the use of milk from or affected animals or through or through the handling of tissues from these animals.

Clinical Features

When it is transmitted to the human, foot and mouth disease manifests by fever, nausea, vomiting, malaise, appearance of ulcerative lesions of oral mucosa and pharynx. Vesicular lesions appear on palms and soles.

Oral Manifestation

Occur at any site, lips, tongue, palate and oropharynx are chiefly affected.

Measles

Acute, contagious, dermatropic viral infection, primarily affecting children, occurring many times in epidemic form caused by paramyxoviridae, which is an RNA virus. Spread of disease occurs by direct contact, or by droplet infection, portal of entry being respiratory tract.

Pathogenesis

Upon invasion of respiratory epithelium it reaches reticulo-endothelial system through blood stream and infect skin, respiratory tract and other organs. Invasion of T-lymphocytes and other increased levels of suppressive cytokines leads to transient supression of cellular immunity. Monocyte is mainly affected. Symptoms show mainly due to the infection with bacteria. Viremia develops, but specific antibodies are not developed before onset of rash.

Cellular immunity plays a major role in host defense against measles.

Clinical Features

Incubation period is 8 to 10 days. Characterised by onset of fever, malaise, cough, conjuctivitis, photophobia, lacrimation and eruptive lesions of skin and oral mucosa. Skin eruptions begin on face, hairline and behind ears, spread to neck, chest and extremities appear as tiny red macules or papules which enlarge and coalesce to form blotchy, discoloured irregular lesions which blanch upon pressure and gradually fade away in 5 to 6 days with fine desquamation.

Oral Manifestations

Oral lesions are prodromal, appear 2 to 3 days before the cutaneous rash are called Koplik's spots.

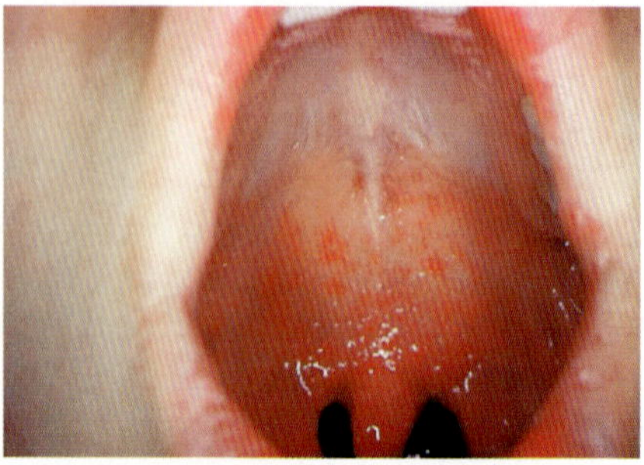

Oral lesions in measles

Immune reaction to the virus in the endothelial cells of dermal capillaries play an important role in development of spots. Also appear on buccal mucosa are small, irregularly shaped flecks surrounded by bright red margin. Palatal and pharyngeal petechiae, generalised inflammation, congestion, swelling, focal ulceration of gingiva, palate and throat may occur.

Complications

Lowers the body immune system may lead to bronchial pneumonia, encephalitis, otitis media, noma.

Control measures: Measles vaccines are available in combinations, e-z strain 5 ml is given subcutaneosly booster doses of MMR is given at 15 to 18 months.

Rubella (German Measles)

Tonsils are enlarged mildly, and red macules appear on palate. Complications may arise during patient in the first trimester of pregnancy offspring has high risk of congenital defects such as blindness, deafness, cardiovascular abnormalities sometimes miscarriage may also occur.

Smallpox

Smallpox is an acute viral disease. It is epidemic in nature and accounted for literally millions of deaths.

On December 9, 1979, the WHO global commission for the certification of smallpox eradication

1. Smallpox eradication has been achieved throughout the world

2. There is no evidence that smallpox will return as an endemic disease.

Clinical Features

Smallpox after an incubation period of 7–10 days manifests itself clinically by the occurrence of high fever, nausea, vomiting chills and head ache. The patient is extremely ill and may become comatose during this period.

The skin lesion begin as small macules and papules which first appear on the face, but rapidly spread to cover much of the body surface. Within a few days the papules develop in to vesicles which themselves eventually become pustular. The pustules are small, elevated, and yellowish green with an inflamed border. They are secondarily infected and occasionally become hemorrhagic, a most serious portent.

Oral Manifestations

Ulceration of oral mucosa and pharynx is rather common, similarly involvement of other mucous membrane such as trachea, esophagus, and vagina also occurs. Multiple vesicles appear much like the cutaneous lesions but, instead of proceeding to the development of pustules, they rupture and form ulcers of nonspecific nature. In some cases the tongue is swollen and painful, makes swallowing difficult.

Molluscum Contagiosum

Molluscum contagiosum is a disease caused by a virus of pox group. The lesion, which only occur on skin or mucosal surfaces are often consider tumor like in nature because of typical localized epithelial proliferation caused by virus.

Clinical Features

The infection is more common in children and adults, and manifest itself as single, or more frequently, multiple discrete elevated nodules, usually occurring on the arms and legs, trunk and face, particularly the eyelids. Its now recognized the disease can be sexually transmitted and lesions of genitalia and pubo-abdominal area also occur with some frequency.

These lesions are hemispheric in shape and usually about 5 mm in diameter with a central umbilication which may be keratinized, and are normal or slightly red in colour. These disease appears to be spread by autoinoculation, by direct contact with affected individual with a reported incubation of 14–50 days. Lesions can occur anywhere in the body, other than palms and soles and may be associated with eczematous rash. HIV-infected individuals are more prone to this lesion. Lesions in HIV-infected patients are atypical in manifestations. Lesions are multiple and grow to a large size resembling carcinomas. The virus replicates in stratum spinosum and forms inclusion bodies which are characteristic and pathognomonic of poxvirus infection called cowdry type.

Oral Manifestations

Mucous membrane involvement, particularly the oral cavity is not common. Cases have been reported, however, such as that of barsh. Oral lesions, which occur most frequently on lips, tongue and buccal mucosa, are similar to those on the skin.

Histological Features

The lesion is quite characteristic, showing thickening and down growth of the epithelium with the formation of large eosinophilic intracytoplasmic inclusion bodies known as Henderson-Paterson inclusions. Measuring approximately 25 microns in diameter. These bodies characteristically accumulate in the crater formed by the distinctive central umbilication of the domed-shaped lesions.

Diagnosis

Smears are prepared from lesion's scrapping or its content and stained with pap stain and Giemsa stains.

Treatment

They are treated by surgical excision or by topical application of wide variety of drugs such as podophyllin and canthradin.

Chickenpox

It is an acute ubiquitous, extremely contagious disease usually occurring in children and characterized by exanthematous vesicular rash. It is most common in winter and spring months.

Etiology

Varicella zoster virus is a DNA virus, which causes two distinct lesions known as chickenpox and a reactivated lesion known as herpes zoster. The incubation period is approximately 2 weeks. The most contagious disease which sooner or later affects everyone in the world. It closely resembles smallpox but is less severe. The mode of transmission is by airborne droplets with port of entry being respiratory tract.

Clinical Features

Characterized by prodormal occurrence of headache, nasopharyngitis and anorexia, followed by maculopapular or vesicular eruptions of skin of low grade fever. These eruptions usually begin in the trunk and spread to involve face and extremities. They occur in successive crops so that vesicles in different stages of formation or resorption may be present. The lesions of skin eventually rupture form a superficial crust and heal by desquamation. Occassionally secondary infection of vesicles results in formation of pustules which may leave small pitting scar upon healing.

Oral Manifestations

Small blister like lesions involve oral mucosa mainly buccal mucosa, tongue, gingiva and palate as well as mucosa of pharynx. These lesions initially slightly raised vesicles with a surrounding erythema, rupture soon after formation form small eroded ulcers with a red margin.

Complications

Secondary infection is common and caused by *Streptococcus* or *Staphylococcus*, CNS involvement leads to encephalitis and transversemyelitis. Perinatal infections are fatal.

Herpes Zoster (Shingles)

Herpes zoster is an acute infectious viral disease of an extremely painful and incapacitating disease which is characterized by inflammation of dorsal root ganglia or extramedullary cranial nerve ganglia, associated with vesicular eruption of skin or mucous membrane in areas supplied by affected sensory nerves. The virus causing the disease is same as that of varicella and both are clinically indistinguishable. Initially a primary infection by V-Z virus causes chickenpox while a recurrent one causes herpes zoster.

Clinical Features

Most common disease of adult life and affects males and females with equal frequency. Initially fever followed by malaise and pain and tenderness along the course of involved sensory nerves usually unilateral. Often trunk is involved. Patient has linear papular or vesicular eruption of skin or mucosa supplied by affected nerve. Triggering factors are trauma, development of malignancy or tumour involvement of dorsal root ganglia, local X-ray radiation.

Oral Manifestations

Unilateral involvement of ophthalmic, maxillary or mandibular nerves. Extreme painful vesicles is seen on buccal mucosa, tongue, uvula, pharynx and larynx which leaves an area of erosion after rupture. These lesions will extend up to midline and stop.

Diagnosis

Distribution of lesion, cytologic smears and characteristic multinucleated giant cells and intranuclear inclusions.

Treatment

Antiviral drug therapy results are promising.

Mumps

Mumps is an acute contagious viral infection characterized by unilateral or bilateral swelling of salivary glands usually parotid while submaxillary and sublingual are sometimes involved but seldom without parotid involvement. Other than parotid it involves nerves meninges, pancreas and gonads. This has an incubation period of 2 to 3 weeks.

Epidemiology

Mumps is transmitted through the respiratory route. It can be isolated from the saliva of infected patient in either 7 days before the onset of parotitis or 9 days after it onsets. Most cases were unrecognized due to absence of parotid swellings. Once infected patients develop lifelong immunity against the disease.

Pathogenesis

Once transmitted through droplet nuclei or saliva or fomites it starts replicating in the respiratory epithelium. It spreads through local lymph nodes and subsequently develops viremia. The affected area shows perivascular and interstitial mononuclear cell infiltrates with edema. Necrosis of acinar and ductal epithelium is seen in salivary glands and germinal epithelium of seminiferous tubules.

Clinical Features

Usually preceded by headache chills moderate fever vomiting and pain below the ear. These symptoms are followed by firm somewhat rubbery or elastic swelling or salivary glands frequently elevating the ear which lasts for 1 week. Pain and tenderness may be severe during the rapid phase of parotid enlargement and swelling reaches its maximum in about 3 days and then gradually subsides. The submandibular gland may also involve separately or in conjugation with parotid. Sublingual involvement is rare. Presternal edema may be present in few cases due to pressure on lymphatics in the neck.

Differential Diagnosis

It should be differentiated from other parotid swellings caused by influenza, parainfluenza 1 and 3, coxsackie, HIV and cytomegalovirus, Sjögren's and pleomorphic adenoma.

Diagnosis

Virus can be isolated from saliva, throat swabs 2 days before or 7 days after onset of parotitis. It can also be isolated from CSF. It can also be confirmed by complement fixation test, hemagglutination inhibition or ELISA. Serum amylase is elevated in both parotitis and pancreatitis.

Complications

Other organs of the body which may be affected include tests, ovaries, pancreas, mammary glands, prostate, epididymis and heart. When mumps involves the adult males orchitis is of great danger which may be unilateral and occasionally causes complete sterility. Meningoencephalitis, deafness, mastoiditis are also occasional complications.

Prognosis

Overall prognosis is good in uncomplicated cases. Death occurs due to CNS or cardiac involvement.

Treatment

Treatment is conservative maintaining hydration and alimentation. Prevention is by vaccination.

HIV Infection (Human Immunodeficiency Virus Infection)

Acquired immunodeficiency syndrome(AIDS) is caused by the Human immunodeficiency virus (HIV) and is characterised by immunosuppression, which leads to spectrum of clinical manifestations. Two genetically distinct populations of viruses known to cause AIDS are HIV1 and HIV2. HIV1 is most common type responsible for AIDS in central Africa and the world, and HIV2 in West Africa and India.

Human Immunodeficiency Virus

The human immunodeficiency virus is a nononcogenic human retrovirus that belongs to the lentivirus group type 3. The virion is an icosehedral structure that contains numerous external spikes formed by two major envelope proteins, the external gp120 and the transmembrane gp41. The HIV glycoprotein antireceptors(gp41 and gp120) make a snug attachment of their specific receptors on a human cell membrane. These receptors are CD4, which are typically found on T4 lymphocytes and other leukocytes.

Mode of HIV Transmission

Sexual transmission is predominant mode of infection worldwide, accounting for more than 75% of all cases of HIV transmission. It is also transmitted through infected body fluids such as blood and blood products, and breast milk. HIV has been demonstrated in oral fluids, but their infectivity appears to be reduced. So, saliva is not a significant route of transmission of HIV infection. Infection can be transmitted vertically from mother to child.

Pathogenesis of HIV Infection

Two major targets of HIV infections are the immune system and central nervous system. A characteristic feature of AIDS is profound immunosuppression of cell mediated immunity. After gaining entry into host, the first step is binding of virion to surface of target cell. This is mediated by binding of envelope gp120 to CD4 surface protein found on most of T helper cells. After binding, the virus become internalised into host cell where it becomes uncoated. The viral RNA is then reverse transcribed into linear double-stranded viral DNA in the cytoplasm of infected cells by reverse transcriptase and then transported to the nucleus, where it is first circularized and then integrated into the host cell. The integrated viral DNA is transcribed into full length RNA by host cell. These RNA transcripts can either be packaged within virus particle, to serve as genome of progeny virus, or they can serve as mRNA for synthesis of viral gene products.

HIV first infects macrophages where it multiplies and is shed, infecting lymphocytes and other cells that possess the CD4 receptor site. Following infection, the virus often enters a dormant stage lasting for 2–15 years.

When HIV is transmitted, fusion with CD4 cells occurs rapidly, and within few days the virus migrates to regional lymph nodes, and then enters the circulatory system. This results in widespread dissemination to the brain and the lymphatic system, causing the 'primary HIV infection' or the 'seroconversion syndrome'.

The most common signs and symptoms of acute infection include fever, fatigue, maculopapular rash, headache, lymphadenopathy, pharyngitis, myalgia, arthralgia, aseptic meningitis, retrorbital pain, weight loss, depression, gastrointestinal distress, night sweats, and oral or genital ulcers. The acute illness may last from a few days to more than 10 weeks, but the duration is usually less than 13 days.

After infection, there is rapid rise in plasma viremia, often to levels in excess of 1 million viral RNA molecules per milliliter. After initial rise in plasma viremia, there is marked reduction from the peak viremia to a steady-state level of viral replication. The decrease in the viral load during acute HIV1 infection is probably due to virus specific immune responses that limit viral replication. There is also a correlation between cytotoxic T-lymphocyte responses to envelope protein and reduction in plasma viral RNA. In addition, soluble factors produced by CD8 cells inhibit HIV1 replication in early stages of acute infection and may thus contribute to the reduction of the viral load. Within 6–12 weeks, antibiotics to HIV are detectable in the blood, and enzyme linked immunosorbent assay(ELISA) and Western blot testing can document seroconversion. At this time, patients are confirmed as HIV positive. The period between viral infection and seroconversion is referred to as 'Window period'. During this period, HIV is present in body fluids and can be transmitted, but the HIV serology may be nonreactive. The disease subsequently progresses to asymptomatic chronic infection with or without persistent generalized lymphadenopathy(PGL). PGL consists of lymphadenopathy for a duration of more than three months and involves two or more extrainguinal sites. Though immune function is normal during this phase, viral replication occurs at rapid rate resulting in destruction of CD4 cells. At this time, the lymphatic tissue serves as major reservoir for HIV, the follicular dentritic cells filter and trap free virus and infected CD4 cells, and viral burden in peripheral blood mononuclear cells is relatively low. With progressive disease, the lymph node architecture is disrupted, and more HIV is released. As more virus is generated and released, more CD4 cells are destroyed, leading to decreased immune function and disease progression.

Clinical Features

The clinical features of HIV infection, range from asymptomatic infection to severe clinical illness and AIDS. Patients generally remain asymptomatic until their CD4 cell count falls below 500 cells/cu.mm, at which time the patient enters the symptomatic HIV infection previously known as 'AIDS related complex', which is characterised by development of opportunistic infections. Without antiretroviral therapy, the progression of HIV from transmission to death is approximately 10–12 years.

Oral Lesions in HIV Infection

Oral Candidiasis

The most common intraoral opportunistic fungal infection strongly associated with AIDS. It has been reported that oral/oesophageal candidiasis, in HIV infected patients, develop full blown AIDS within two years.

Pseudomembranous candidiasis—Appears as white to yellowish white plaques, easily scrapable exposing red areas.

Erythematous candidiasis—Clinically seen as red lesions, commonly located on dorsum of tongue, palate, and buccal mucosa.

Hyperplastic candidiasis—These lesions are characterised by white plaques which cannot be removed by scraping. Biopsy demonstrates fungal hyphae in keratinised layers of epithelium.

Angular cheilitis—Erythema and/or fissuring and/or scaling of angles of mouth clinically characterize this lesion.

Periodontal Lesions

Linear gingival erythema—Erythematous band of marginal gingiva with either diffuse or punctate erythema of attached gingiva.

Necrotising periodontal diseases are subclassified as necrotising ulcerative gingivitis (NUG), necrotising ulcerative periodontitis (NUP) and necrotising stomatitis (NS). NUG involves destruction of one or more interdental papilla and limited to marginal gingiva.

NUP extends beyond papillae and marginal gingiva causing loss of periodontal attachment, and exposing the bone. When necrosis extends beyond the periodontium into mucosa and osseous tissue it results in NS.

Chronic periodontitis-increased rate of attachment loss or exacerbated periodontitis.

Herpes Simplex Virus Infection

Herpes labialis—seen as vesicles on lip and adjacent facial skin which rapidly break down to produce shallow ulcers.

Herpes zoster virus infection—is a recurrent viral infection seen in AIDS patient and presents clinical course, which is more severe in morbidity. In AIDS patient herpes zoster begins as a unilateral cluster of vesicles and ulcers in a classical dermatome distribution

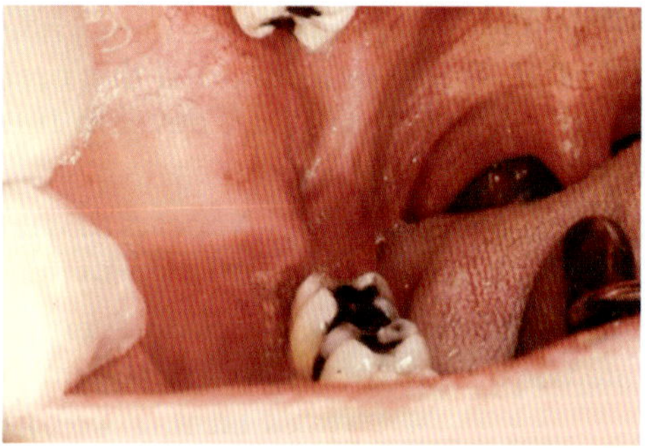

Herpes labialis

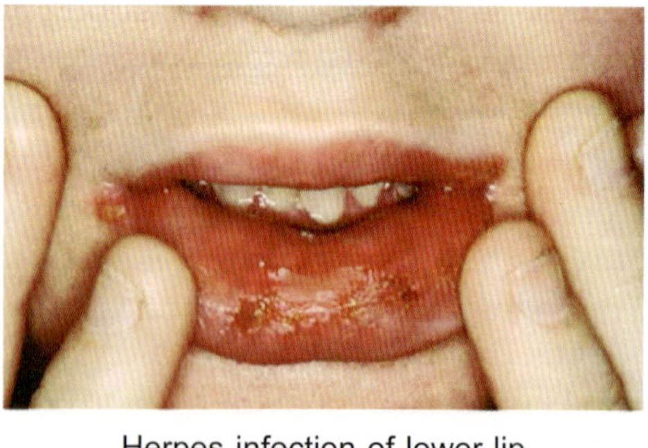

Herpes infection of lower lip

but subsequently extends beyond the dermatomal boundary and heals by scarring.

Oral Hairy Leukoplakia

Epstein-Barr virus is associated with these lesions. Basal epithelial cells of lateral margin of tongue normally harbor latent EBV and significant diminution of Langerhans cells by HIV, permits reactivation of EBV with subsequent epithelial hyperplasia. OHL frequently appears bilaterally on lateral borders of tongue as

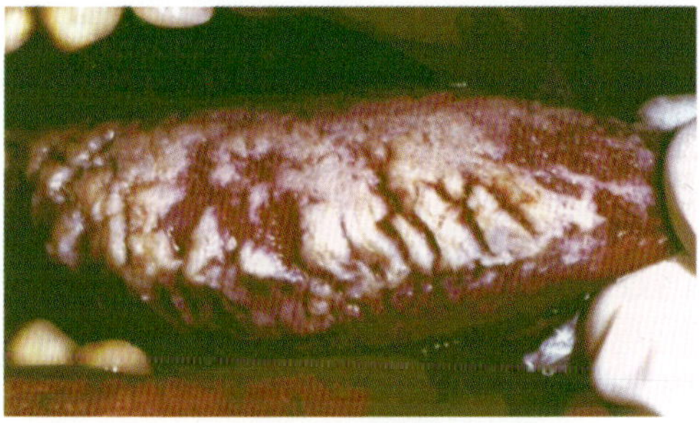

Hairy leukoplakia

painless, faint white vertical streaks or thickened and furrowed areas with shaggy keratotic surface with vertical striations imparting corrugated appearance.

Kaposi's Sarcoma

Kaposi's sarcoma is multifocal neoplasm of vascular endothelial origin. Human herpes virus type 8 is involved in pathogenesis of Kaposi's sarcoma. Commonly affected sites include palate, gingiva, tongue, and oropharynx or skin. The clinical appearance of oral Kaposi's sarcoma can be macular, nodular, or raised and ulcerated, color of which can range from red to purple. Early lesion tend to be flat, red and asymptomatic, with color becoming darker as lesion ages. With secondary infection or trauma becomes symptomatic.

Aphthous Ulcers

Aphthous ulcer occurs as recurrent, round, shallow, painful ulcers of variable size and duration that are typically found on

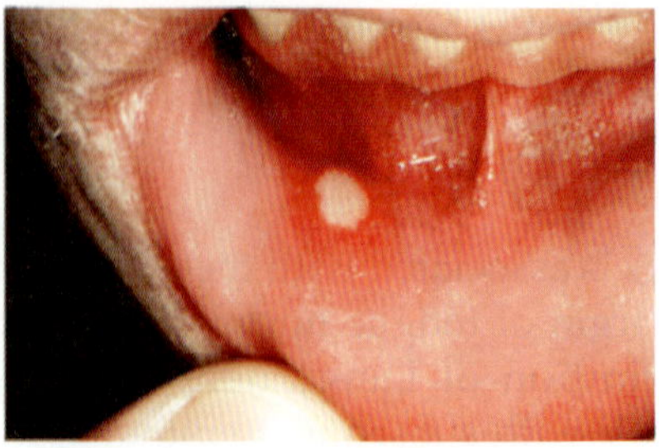

Aphthous of lower lip

nonkeratinized oral mucosa. Oral ulcers in HIV patients are large and more extensive measuring more than 2 cm in diameter with irregular borders.

Oral Squamous Cell Carcinoma

Oral squamous cell carcinoma has been reported in AIDS patients. Impaired immunosurveillance and an increased chance of human papilloma virus infection are few suggested causes.

Molluscum Contagiosum

Molluscum contagiosum is an infection of skin caused by pox virus. It presents as shiny, white and hemispherical skin colored dome-shaped papules that often demonstrate a central depressed crater. In AIDS patient numerous lesions may present which have no tendency to undergo spontaneous resolution and some grow to large size.

Thrombocytopenic Purpura

Thrombocytopenic purpura in HIV disease is characterized by reduced production of platelets due to drugs, malnutrition, immunologic alterations, microbial invasion, or due to course of HIV disease. Oral TP is characterized by pin point petechiae following minor trauma and even mastication

HIV associated Salivary Gland Disease

HIV associated salivary gland disease is heterogeneous group of signs and symptoms that vary from person to person and as the

HIV infection progresses in the same individual. The main clinical sign is bilateral parotid enlargement and associated with cervical lymphadenopathy. Quantitative and qualitative xerostomia is experienced by most patients. The causes of xerostomia include drugs (antiretroviral, antifungal, chemotherapeutics, antihistamines, mood altering drugs, multivitamins or as a part of the progression of HIV disease. The symptoms include dryness of mouth, predisposition to fungal diseases, dental caries, and infections.

Non-Hodgkin's Lymphoma

Aggressive B cell lymphoma is second most common cancer associated with HIV. Epstein-Barr virus is associated with non-Hodgkin's lymphoma.

Hyperpigmentation

Oral hyperpigmentation may occur suddenly in HIV infected individuals and has been attributed to several medications including ketoconazole and zidovudine taken by AIDS patients. Adrenocortical destruction has been reported from several infections associated with HIV disease.

Diagnosis of HIV

Virus based tests

1. *Viral culture:* It directly detects the virus and hence is highly specific. It is positive, even during window period and during terminal phase when the viral load is high and when the ELISA may be negative.
2. *PCR (Polymerase chain reaction):* PCR for DNA of pox virus present in infected host cells and PCR for HIV RNA from plasma, which detect the free virus present in plasma. PCR can detect as low as 20 copies/ml of plasma and is highly sensitive and specific test. It is very useful for early diagnosis of HIV infection in newborn.
3. *P24 antigen detection:* It is detectable during the window period and the late phase of infections when the virus is replicating fast. It is often not detectable at other times as viral replication

is low. Hence this test is not sensitive enough to be of use for routine diagnosis. It is useful in blood banks where P24 antigen, along with ELISA, can shorten the window period of diagnosis to less than two weeks.

Anti-HIV Antibody Tests

1. *Enzyme-linked immunosorbent assay (ELISA):* ELISA is more popular and widely done test. ELISA have nearly 99% sensitivity and specificity, especially on repeated testing. In west it is mandatory to do a western blot if ELISA is positive. In India and according to WHO, it is not necessary to do a western blot if ELISA is positive. Instead, one should confirm it by repeating one or two more ELISA tests, using different kits on repeated blood samples. So, if two or preferably three ELISA are positive, a person is confirmed as HIV infected. ELISA cannot diagnosis HIV infection in newborn. ELISA is negative during window period.
2. *Western blot analysis:* This detects specific antibodies and shows them as separate bands on gels. Hence this is more specific than ELISA. It is interpreted as positive if atleast two or three bands, e.g. P24, gp41, gp 120/160 bands, are positive. If one band is positive, it is labeled as indeterminate.

Immunological tests and surrogate markers—$CD4^+T$ cell count, $CD4^+T$ cell% and CD4/CD8 ratio are tests done to determine immune status. A gradual decline in $CD4^+T$ cell count suggests disease progression.

Salivary Tests

The fluids obtained from the oral cavity are saliva and gingival crevicular fluid. The immunoglobulin content of oral fluid is similar to that of blood, but their levels are lower. However, the use of an HIV IgG antibody capture assay, designed specifically for testing oral fluids, has produced encouraging results.

6 Uncommon Lesions with Viral Etiology

Papilloma

The oral squamous papilloma is associated with papilloma viruses, the one commonly incriminated as causative in skin warts. The HPV types 6–11 commonly associated with squamous papillomas. Even though all HPV lesions are infective, the squamous papilloma appears to have an extra an extremely low virulence and infectivity rate. It does not seem to be contagious.

Clinical Features

The papilloma is an exophytic growth made up of numerous small finger like projections which result in a lesion with rough ended verrucous or cauliflower surface. It is most commonly found in tongue, lips, buccal mucosa, gingival and palate.

The common wart or verrucous vulgaris is a frequent tumour of the skin analogous to the oral papilloma. It is uncommon on oral mucous membranes but extremely common on the skin.

Oral Papilloma

It enlarges rapidly to its maximum size achieving more than 5 mm in greatest diameter. Verucca vulgaris is contagious and capable

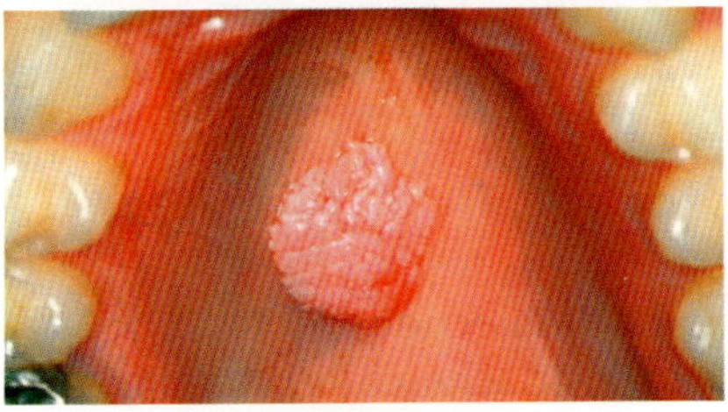

Oral papilloma

of spread to other parts of an affected person's skin or membranes by way of auto inoculation. Papilloma like or papillomatous lesions as well as pebby lesions and fibromas of various sites in the oral cavity are recognized as one of the many manifestations of the multiple hamartoma and neoplasia syndrome (Cowden's syndrome).

Histological Features

Microscopically papilloma consist of long, thin, finger like projections extending above the surface of the mucosa, each made up of continuous layer of stratified squamous epithelium and containing a thin, central connective tissue core.

Treatment and Prognosis

Papilloma is treated by surgical excision, including the base of the mucosa into which the peduncle or stalk inserts. Removal should never be accomplished by an insertion through the peduncle. If it is properly excised recurrence is rare.

Intraoral verruca vulgaris is also treated effectively by conservative surgical excision or curettage but liquid nitrogen cryotherapy and topical application of keratinolytic agents and are also effective.

Infectious Mononucleosis

Infectious mononucleosis also known as glandular fever.

Occurs in children and young adults. Since 1800s it has been recognized as clinical syndrome consisting of fever, pharyngitis, and adenopathy. EB virus is transmitted via intimate contact with body secretions primarily oropharyngeal secretions and one important means is thought to be 'deep kissing' or intimate oral exchange of saliva. For this reason the condition referred to as 'Kissing disease'.

Clinical Features

Seen in epidemic form, infectious mononucleosis is characterized by fever, sore throat, headache, chills, cough, nausea, vomiting and lymphadenopathy (bilateral and symmetrical). Splenomegaly and hepatitis occur frequently. Most patients with EBV can be asymptomatic.

Cervical lymph nodes usually exhibit enlargement, followed by nodes of axilla and groin. Pharyngitis and tonsillitis are common and skin rashes has occasionally reported.

Majority of cases of children appear to be asymptomatic. Peak incidence occurs in 15 to 20 years age group.

Oral Manifestations

There are no specific manifestations in oral cavity but secondary lesions do occur. Chiefly consisting of acute gingivitis and stomatitis, appearance of white or grey membrane in various areas, palatal petechiae and occasional oral ulcers.

Petechial hemorrhages of soft palate near junction with hard palate is an early manifestation of infectious mononucleosis. Pinpoint petechiae appear in dozen to several hundred in few days and fade away gradually after staying for 3–11 days.

Laboratory Findings

Patient exhibits atypical lymphocytes in circulating blood, as well as antibodies to EB virus. Increased neutrophil antibody titer is present only in small minority of children with the disease. The normal titer of agglutinins and hemolysins in human blood against sheep erythrocytes does not exceed 1:8. In infectious mononucleosis titer may rise to 1:4096. This is referred to as Paul-Bunnell test and is both characteristic and pathognomonic of the disease. Patient has more than 50% lymphocytosis of which 10% are more are the 'atypical' forms. These 'atypical' forms consist of either oval, horse-shoe-shaped or indented nuclei with dense, irregular nuclear chromatin and a basophilic. Foamy or vacuolated cytoplasm. A thrombocytopenia is also present in some patients.

Treatment

There is no specific treatment for infectious mononucleosis. Antibiotics have been used without great success. Bed rest and adequate diet are of great benefit. Disease runs its course in two to four weeks and there seldom are complications.

Leukemia

Leukemia is characterized by progressive over production of leukocyte which appear in circulating blood in immature form. Proliferation occurs in uncoordinated way and is considered as malignant neoplasm, particularly since the disease is fatal.

Any of white blood cells may be involved and disease may be classified as follows:

1. Lymphoid (lymphoblastic, lymphocytic) leukemia involving lymphocytic series.
2. Myeloid (myelogenous) leukemia-involving myeloid series erythrocytes, granulocytes, monocytes and platelets.

Acute form of leukemia is one in which survival is less than 6 months and chronic form implies survival of over one year.

Etiology

Unknown etiology. Viruses have been suspected to be etiologic factor. Epstein-Barr virus has been implicated as being most likely leukomogenic virus. The importance of cofactors or predisposing factors such as genetics, age, hormones, immune competence and stress, must all be considered in determining the susceptibility to tumor development of an individual with an oncogenic virus.

Clinical Features

Acute leukemia common in children and young adults, while chronic leukemias more common in middle aged or older.

Acute Leukemia

Development of acute leukemia is sudden characterized by weakness, fever, headache, generalized swelling of lymph nodes, petechial or ecchymotic hemorrhages in skin and mucous membranes and evidence of anemia. Lymphadenopathy is often first sign of disease. Numerous organs such as spleen, liver and kidney, become enlarged, owing to leukemic infiltration, especially in long duration cases.

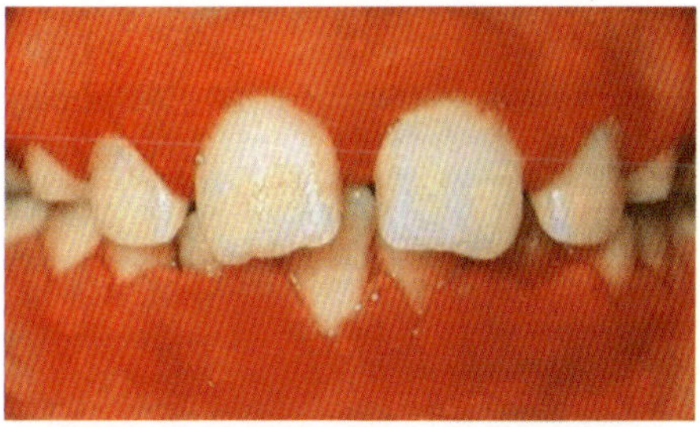

Acute leukemic gingival changes

Chronic Leukemia

Chronic leukemia develops so insidiously that the disease may be present for months or even several years before the symptoms lead to discovery.

Patient exhibit features such as anemic pallor. Lymph node enlargement is common in chronic lymphatic leukemia. Enlargement of salivary glands and tonsils occur, owing to leukemic infiltration and thus results in xerostomia. Petechiae or ecchymoses of skin is seen. In other instances there may be leukemids: papules, pustules, bullae, areas of pigmentation, herpes zoster, itching and burning sensations or variety of other disturbances. Finally nodular lesions composed of leukemic cells may occur on skin. Destructive lesions of bone are reported in some cases of chronic leukemia and these may result in pathologic bone fractures or osteomyelitis.

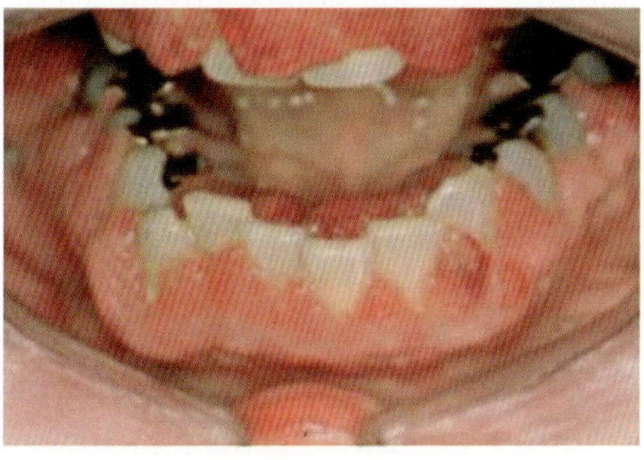

Inflammed gingival due to leukemia

Laboratory Findings

Hematologic examination constitutes the basis for final diagnosis of any leukemia.

Acute Leukemia

Anemia and thrombocytopenia are both characteristic of acute leukemia. As a result both bleeding time and coagulation time are prolonged. Leukocyte count subnormal in early stages and rises in terminal stages to 1,00,000 or more cells per cubic mm of blood. In myeloid type predominant cells resembles myeloblast or undifferentiated myelocyte.

Chronic Leukemia

Anemia and thrombocytopenia are common in chronic form of leukemia. Leukocyte count over 5,00,000 cells per cubic mm. On other hand very low level count also occurs.

Oral Manifestations

Oral lesions occur in both acute and chronic forms of all types of leukemia: myeloid, lymphoid and monocytic.

Primary clinical manifestations may consist of gingivitis, gingival hyperplasia, hemorrhage, petechiae and ulceration of the mucosa. Gingival hyperplasia, which may be constant features. In severe cases the teeth may be hidden. Gingivae are boggy, edematous and deep red. They bleed easily. The gingival bleeding is due to leukemic infiltration in areas of mild chronic irritation. Purpuric lesions of oral mucosa analogous to cutaneous ecchymoses may also be seen. Gingival hemorrhage occurs commonly is due to ulceration of sulcular epithelium necrosis of underlying tissue. Since normal leukocyte distribution is disturbed normal inflammatory response is disturbed. Rapid loosening of teeth due to necrosis of periodontal ligament and destruction of alveolar bone occurs.

Treatment

Wide array of chemotherapeutic drugs, radiation therapy, and corticosteroids under certain circumstances offer prolonged remissions and apparent cures in atleast some forms of disease.

Nasopharyngeal Carcinoma

A tumor of the nasopharynx involving squamous epithelium. The evidence that EBV is involved in pathogenesis is based on detection of multiple copies of the EBV genome can be detected in malignant cells of 100% of undifferentiated nasopharyngeal carcinoma.

Clinical Features

Environment as well as genetic factors is involved. HLA haplotypes and environmental factors such consumption of salted fish and food containing nitrosamines are associated with NPC. Tumor commonly arises in posterior wall of nasopharynx and metastasize to local lymph nodes. Common presentation is bilateral involvement of glands in neck. Symptoms of invasion by primary tumor, e.g. nasal obstruction, post-nasal discharge, epistaxis, partial deafness and cranial nerve palsies.

Histologic Features

Three types of NPC are recognized on histologic appearance:
- Well-differentiated squamous cell carcinoma
- Non-keratinizing carcinoma
- Undifferentiated carcinoma.

Treatment

NPC is difficult to treat surgically because of early metastasis to regional metastasis. Tumor is resistant to chemotherapy and radiotherapy is treatment of choice. Tumor usually presents late, the prognosis is poor with 5-year-survival rate of 20%. It is possible to prevent the development of NPC with the use of EBV vaccine at an early age.

Burkitt s Lymphoma (African jaw lymphoma)

A tumor peculiar to the children of tropical central Africa reported by Denis Parsons Burkitt which was later named after him. Burkitt's lymphoma occurs endemically in parts of Africa and Papua.

It is low grade B-cell neoplasm and has two major forms:
- The endemic (African) form

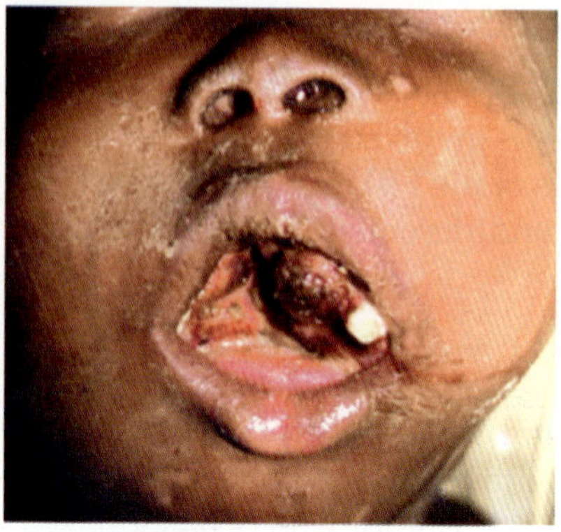

Burkitt's lymphoma

- The non-endemic form
- Burkitt's lymphoma is childhood tumor but seen in adults too. It is one of the fastest growing malignancies in human, with a very high growth fraction.

Clinical Features

African form most often involves the maxilla and mandible. The involvement of abdominal organs such as kidneys, ovaries, or retroperitoneal structures is slightly less common. The sporadic form usually involves abdominal organs, with most common involvement of distal ileum, cecum, or mesentery, and less common involvement of other abdominal organs, pelvic organs, and facial bones.

The exact cause and mechanism of Burkitt's lymphoma is not known. Epstein-Barr virus is closely associated with African form of Burkitt's lymphoma. Some have postulated that, immunosuppression caused by coexistent malaria or another infection, the host is unable to generate adequate T-lymphocyte response (EBV specific cytotoxic T cells) against B cells that are infected latently with EBV. This subsequently results in excessive B cell proliferation. In African form of Burkitt's lymphoma, patients most often present with swelling of lymph nodes, which are nontender and rapidly growing, in neck or below the jaw. Abdominal presentation is less common.

The patients with sporadic form of Burkitt's lymphoma most commonly present with abdominal tumors causing swelling and pain in affected area. Some patients present with symptoms of bowel obstruction secondary to an ilealcecal intussusception caused by tumor growth. Because of rapid growth of Burkitt's tumor, patients may quickly manifest significant metabolic derangement and renal function impairment. Less common presentations of Burkitt's lymphoma include an epidural mass, skin nodules, CNS involvement, and bone marrow involvement. Major signs of Burkitt's lymphoma include a soft tissue mass associated with the involvement of the jaws or other facial bones, enlarged cervical lymph nodes, abdominal masses, and ascites.

Histologic Features

Burkitt's lymphoma is monoclonal proliferation of B lymphocytes characterised by small noncleaved cells that are uniform in appearance and that produce a diffuse pattern of tissue involvement. Burkitt cells are homogenous in size and shape, with round to oval nuclei and slightly coarse chromatin, with multiple nucleoli and with intensely basophilic vacuolated cytoplasm that contains neutral fat. A characteristic 'starry sky' appearance is imparted by scattered macrophages with an abundant clear cytoplasm, often containing phagocytic cellular debris.

Treatment

Before advent of specific therapies, children with Burkitt's lymphoma died rapidly. With combination chemotherapy and CNS prophylaxis (intrathecal chemotherapy), the survival rate is now atleast 60%. Those with bone marrow and CNS involvement have a poor prognosis.

Hodgkin's Lymphoma

(Hodgkin's disease, malignant lymphoma)

Hodgkin's disease is considered as one of two main types of malignant lymphomas. Etiology is unknown. Infectious agents, especially the Epstein-Barr virus may be involved in pathogenesis. Patients with HIV infection have a higher incidence of Hogkin's disease compared to the population without HIV infection.

Clinical Features

Age specific incidence rates have a bimodal distribution in both genders, peaking in young adults and older individuals. The first manifestation in majority of cases is painless enlargement of one or more cervical lymph nodes. Palpable painless lymphadenopathy occurs in cervical areas, axilla, and less commonly inguinal area and waldeyer ring or occipital nodes. The nodes are usually firm, rubbery in consistency, and the overlying skin is normal.

Constitutional symptoms such as unexplained weight loss, fever, night sweats are present in 40% patients. Chest pain, cough, and shortness of breath may be present due to a large mediastinal mass or lung involvement. Rarely, hemoptysis is observed. Patient may present with pruritis or intermittent fever. Back, abdomen or bone pain may occur rarely due to spleenomegaly, hepatomegaly, pressure from enlarged lymph nodes, involvement of bone or vertebrae.

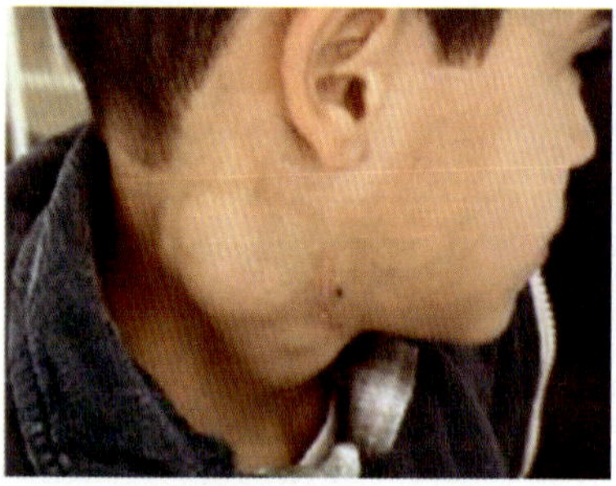

Hodgkin's lymphoma

Oral Manifestations

Oral cavity could be involved secondarily. A case of Hodgkin's disease involving secondarily mandible and overlying alveolar mucosa has been reported.

Characteristics of Typical Reed-Sternberg Cell

Characteristic malignant cells of Hodgkin's diseases are large cells known as Reed-Sternberg cells (typical and variant), 20–50 micrometers in diameter, abundant, amphophilic, finely granular/homogeneous cytoplasm, two mirror image nuclei (owl

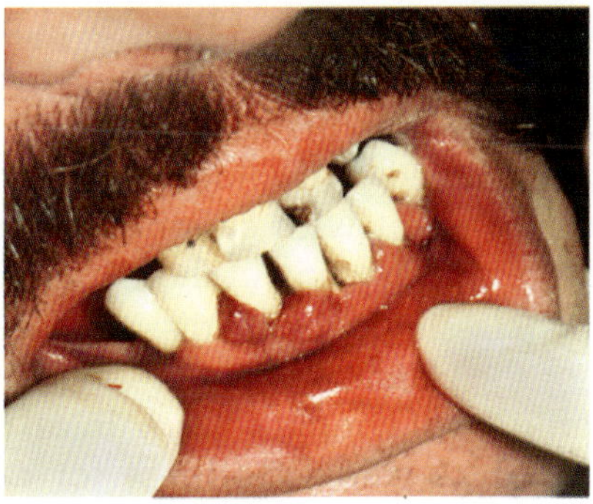

Oral involvement of Hodgkin's

eyes) each with an eosinophilic nucleolus and a thick nuclear membrane (chromatin is distributed at cell periphery). One or two percent of these malignant cells are admixed within a reactive cell infiltrate composed of variable proportions of lymphocytes, histiocytes, eosinophils, and plasma cells. The Reed-Sternberg cells are identified as large binucleated cells with prominent nucleoli and an unusual CD15+, CD30+ immunophenotype. In 50% cases, the Reed-Sternberg cells are infected by the Epstein-Barr virus. There are 5 recognized histologic types of Hodgkin's disease which form the basis for its classification.

Histologic Features

1. Nodular sclerosis Hodgkin's disease–comprises 60–80% of all cases. The morphology shows nodular pattern. The broad bands of fibrosis divide the nodes into nodules. The capsule is thickened. The characteristic cell is lacunar-type Reed-Sternberg cell,which has monolobulated or multilobated nucleus and a small nucleolus with abundant and pale cytoplasm. Usually involves the mediastinum and other supradiaphragmatic sites.
2. Mixed-cellularity Hodgkin's disease—comprises 15–30%. Histologically the infiltrate is usually diffuse. RS cells are of the classic type (large, with bilobate, double or multiple nuclei, and large eosinophilic inclusion like nucleolus). Patients with this histology typically have advanced stage disease with systemic symptoms and immunodeficiency.

3. Lymphocyte-depleted Hodgkin's disease–makes up less than 1%. The infiltrate is diffuse and often appears hypocellular. Large numbers of RS cells and Bizarre sarcomatous variants are present. It is associated with older age and HIV positivity. Patients usually present with advanced-stage disease. EBV proteins are expressed in many of these tumors.
4. Lymphocyte-rich classic Hodgkin's disease-comprises 5%. In this type, RS cells of the classic or lacunar type are observed, with background infiltrate of lymphocytes. Clinical presentation and survival patterns are similar to those for mixed-cellularity Hodgkin's disease.
5. Nodular lymphocyte-predominent Hodgkin's disease-constitutes 5%. RS cells are not observed or appear infrequently. Instead, a variant of RS cells, the lymphocytic and histiocytic cells, or 'pop corn' cells (their nuclei resemble an exploded kernel of corn), are seen within a background of inflammatory cells, predominantly benign lymphocytes.

Treatment

Proper treatment of Hodgkin's disease can lead to long-term remission and even cure. Radiation therapy and combination chemotherapy shown to be effective. The most important prognostic determinants are the histologic type and the clinical stage of disease. Tht lymphocyte predominant type has the most favourable prognosis, followed by nodular sclerosis, mixed cellularity and lymphocyte depletion. Localised disease has much better prognosis than disseminated disease.

Oral Hairy Leukoplakia

Oral hairy leukoplakia was first reported by Greenspan and coworkers on the lateral margin of tongue among young homosexual males. The term hairy was given because of corrugated surface of the epithelium. The prevalence among AIDS patients and in certain immunocompromised HIV seronegative patients seen. The association of these lesions with Epstein-Barr virus has been demonstrated. Its has been hypothesized that basal epithelial cells of lateral margin of the tongue normally harbor latent EBV and significant diminution of Langerhans cells by HIV, in affected

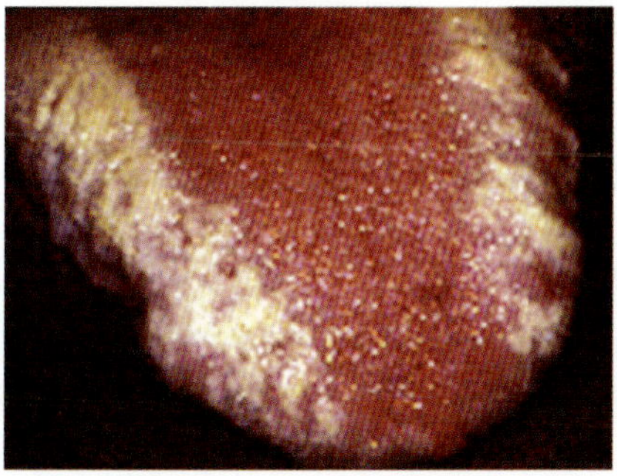

Oral hairy leukoplakia

site, permits reactivation of EBV with subsequent epithelial hyperplasia. Oral hairy leukoplakia is predictive indicators for subsequent development of AIDS in HIV seropositive patients.

Clinical Features

OHL appears bilaterally on lateral borders of the tongue as painless, faint white vertical streaks or thickened and furrowed areas with shaggy keratotic surface with vertical striations imparting a corrugated appearance. The floor of mouth, buccal mucosa, tonsillar region, soft palate, and posterior portion of pharynx are other areas involved less frequent.

Histologic Features

Lesion shows features of epithelial hyperplasia with acanthosis and hyperkeratosis, which produce surface corrugations. A remarkable feature of acanthotic epithelium is presence of varying numbers of lightly stained, swollen, balloon cells in the upper prickle cell layer. Superficial epithelial cells show dense aggregates of nuclear chromatin material marginated along the nuclear membrane known as nuclear beading.

Management

Oral hairy leukoplakia is usually asymptomatic, but occasional slight soreness, discomfort, and unsightly appearance of the lesion warrant treatment. Since oral hairy leukoplakia cases reveal candidiasis, topical antifungal agents nystatin and clotrimazole and systemic agents like ketoconozole and fluconazole are employed. Antiviral agents like cyclovir, desciclovir, ganciclovir, result in clinical improvement. However, discontinuation of therapy results in recurrence of the lesion, if immunosuppression persists.

Warthin s Tumor

Warthin's tumor is also known as papillary cystadenoma lymphomatosum, adenolymphoma. It is second common tumor in salivary glands. This unusual type of salivary gland tumor occurs almost exclusively in the parotid gland, although occasional cases have been reported in submaxillary gland. The intraoral accessory salivary glands are rarely affected.

Histogenesis

The tumor arises in salivary gland tissue entrapped within paraparotid or intraparotid lymph nodes during embryogenesis. Another suggestion is Warthin's tumor is most likely a delayed hypersensitivity disease, the lymphocytes being an immune reaction to the salivary ducts which undergo oncocytic change. A strong association between development of this tumor and smoking is documented. Epstein-Barr virus has also been implicated in the pathogenesis of this tumor.

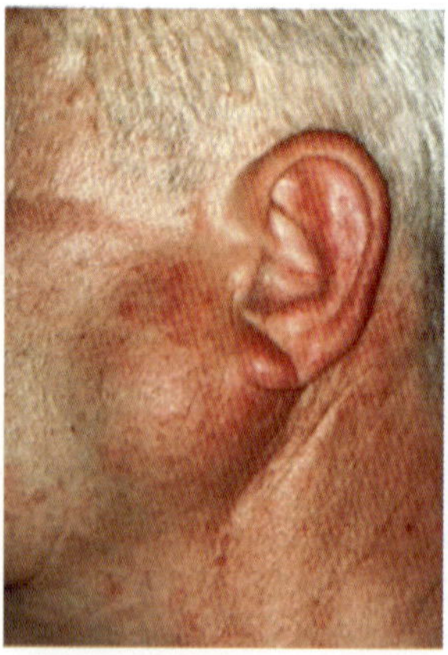

Warthin's tumor showing extraoral swelling

Clinical Features

Commonly presents in the sixth and seventh decades. The tumor is generally superficial, lying just beneath the parotid capsule or protruding through it. It is not painful, is firm to palpation and is clinically indistinguished from other benign lesions of parotid gland.

Histologic Features

Macroscopic features—It is smooth, somewhat soft parotid mass is well encapsulated when located in the parotid. The tumor contains variable number of cysts that contain clear fluid. Areas of focal hemorrhage may also be seen.

Microscopic Features

This tumor is made up of histologic components–epithelial and lymphoid tissue. The lesion is adenoma exhibiting cyst formation, with papillary projections into cystic spaces and a lymphoid matrix showing germinal centers. The cysts are lined by papillary projections of bilayered oncocytic epithelium. The inner cells are tall columnar with finely granular and eosinophilic cytoplasm due to presence of mitochondria and slightly hyperchromatic nuclei. The outer layer cells are oncocytic triangular and occasionally fusiform basaloid cells. There is frequently an eosinophilic coagulum present within the cystic spaces, which appears as a chocolate-coloured fluid in the gross specimen. The abundant lymphoid component may represent the normal lymphoid tissue

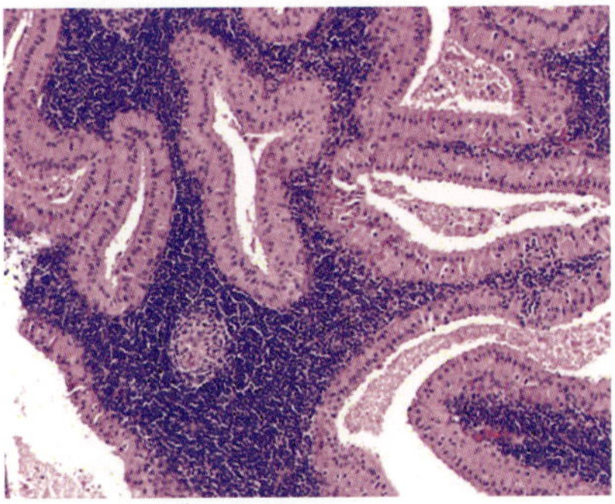

Bilayered oncocytic epithelium

of the lymph node within which the tumor developed or it may actually represent a reactive cellular infiltrate which involves both humoral and cell mediated mechanisms.

Treatment

The accepted treatment is surgical excision. This can almost invariably be accomplished without injury to facial nerve, particularly lesion is usually small and superficial. These tumors are well encapsulated and seldom recur after removal.

7

Fungal Infections

Histoplasmosis

Histoplasmosis is a fungal infection caused by *Histoplasma capsulatum*. It is usually acquired by inhalation of dust containing spores of the fungus, the contamination probably occurring from excreta of birds, such as pigeons, starlings and blackbirds.

Clinical Features

The disease is characterized by a chronic low grade fever, productive cough, splenomegaly and lymphadenopathy, since the

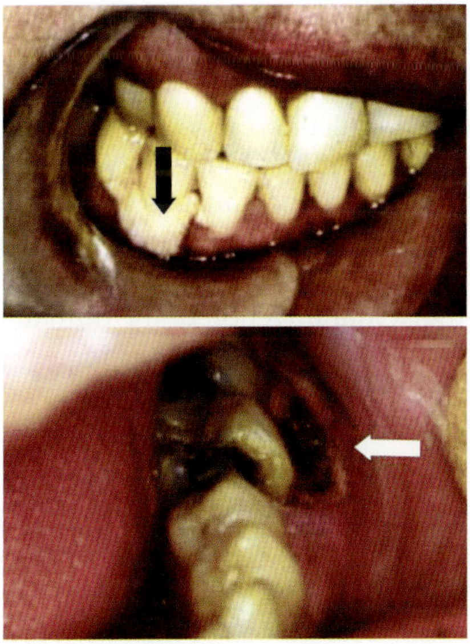

Histoplasmosis

organism have a special predilection for the reticuloendothelial system and chiefly involve the spleen, liver, lymph nodes, and the bone marrow. The infection by this organism may be extremely mild manifesting only local lesions such as subcutaneous nodules or suppurative arthritis and may produce no more serious effects than a positive histoplasmin skin reaction or calcified pulmonary nodules similar to those seen in tuberculosis.

Oral Manifestation

They appear as nodular, ulcerative or vegetative lesions on the buccal mucosa, gingival, tongue, palate or lips. The ulcerated areas are usually covered by a nonspecific gray membrane and are indurate. The organisms may be demonstrated in tissues sections in many, but not all, cases. Thus it is wise in suspected cases to preserve a piece of tissues at the time of biopsy for microbiologic examination.

Treatment

Pulmonary histoplasmosis usually, resolves spontaneously, while severe forms of the disease are usually treated by amphotericin B.

Cryptococcosis

It is a chronic fungal infection caused by *Cryptococcus neoformans* and *Cryptococcus bacillispora* and may present wide spread lesions in the skin, oral mucosa, subcutaneous tissues, lungs, joints, and particularly the meninges. The organisms are frequently found on the skin of healthy persons. The organism appears to be harboured by pigeons, but the actual infection of the human probably results from inhalation of airborne microorganism.

Clinical Features

The first evidence of infection by these organisms is the presence of skin lesions from which blood stream dissemination to other parts of the body frequently occurs. The skin lesions appear as multiple brown papules which ultimately ulcerate, the clinical picture is not specific. The lesions of the lungs produce symptoms of a nonspecific pneumonitis while the meningeal lesions produce a variety of neurologic signs and symptoms generally associated with increased intracranial pressure.

Oral Manifestation

The oral lesions appear as simply nonspecific single or multiple ulcers which in a patient known to have leukemia may be mistaken for the wide spread ulceration often seen in leukemic patients as a result of their inability to react to a mild nonspecific bacterial infection.

Treatment and Prognosis

The use of amphotericin B has been found to give good results. The ultimate prognosis of the patient is variable however and especially dependent upon the sites of involvement.

Sporotrichosis

Sporotrichosis is a fungal infection caused by sporotrichum in which the portal of entry is not clearly understood.

It has been reported to occur after,

1. Exposure to a wide variety of animals both domestic and wild
2. Accidental injury from the thorns of some plants or bushes and
3. Accidental laboratory or clinical inoculation of hospital workers.

Clinical Features

The most common lesions of sporotrichosis involve the skin, subcutaneous tissues and oral, nasal, and pharyngeal mucosa, although disseminated visceral involvement occasionally occurs. The skin lesions often describe as sporotrichotic 'chancres' appear at the site of inoculation as firm, red to purple nodules, which soon ulcerates.

Oral Manifestations

Nonspecific ulceration of the oral, nasal and pharyngeal mucosa also occurs in the disease usually associated with regional lymphadenopathy. The lesions are described as healing by soft pliable scars even though the organisms may still be present in the tissues.

Treatment and Prognosis

There is no specific treatment for the disease most antibiotics being ineffective but the prognosis is generally good although chronic repetitive remissions and relapses are common. The common pulmonary form of the disease is often fatal.

Rhinosporidosis

It is a chronic granulomatous disease caused by a fungus called *Rhinosporidium seeberi* which affects chiefly the oropharynx and nasopharynx as well as larynx, skin, eyes, and genital mucosa.

Clinical Features

Nasal mucosa is the most common site involved. Lesions appear as small verucae or warts which ultimately become pedunculated. Genital lesions resemble condylomas.

Oral Manifestations

The oropharyngeal lesions are often accompanied by mucoid discharge and appear as soft red polypoid growth of a tumour like nature which spread to the pharynx and larynx. The lesions are vascular and bleed regularly. Though any intraoral site may be involved soft palate appears to be the most frequent site.

Treatment

Surgical removal of the growth is recommended as a treatment of choice.

Candidiasis

Candidiasis is caused by yeast-like fungus *Candida albicans*. *Candida* exists in three forms namely pseudohyphae, yeast, and chlamydospore forms. It reproduces by asexual budding and forms pseudohyphae It is relatively common inhabitant of oral cavity, gastrointestinal track and vagina, of clinically normal persons. There must be actual penetration of tissues although such invasion is usually superficial and occurs only under certain circumstances. This disease is said to be most opportunistic infection in the world occurrence has increased remarkably since the prevalent use of antibiotics which destroy the normally inhibitory bacterial flora and immunosuppressive drugs particularly corticosteroids and cytotoxic drugs. Oral candidiasis or thrush usually remains a localized disease but on occasion it may show extension to the pharynx or even to lungs often with a fatal outcome.

Classification of Oral Candidiasis

Acute

1. Acute pseudomembranous oral candidiasis
2. Acute atrophic oral candidiasis

Chronic

1. Chronic hyperplastic oral candidiasis
2. Chronic mucocutaneous oral candidiasis
   - A. Chronic familial mucocutaneous candidiasis
   - B. Chronic localized mucocutaneous candidiasis
   - C. Chronic diffuse mucocutaneous candidiasis
   - D. Candidiasis endocrinopathy syndrome
3. Chronic atrophic oral candidiasis.

Clinical Features

It has variety of clinical manifestation making the diagnosis some times difficult. It may range from mild superficial mucosal involvement to severe fatal disseminated form seen in into two major categories.

1. Mucocutaneous candidiasis
2. Systemic candidiasis.

Acute Pseudomembranous Candidiasis

It is one of the most common forms of the disease especially prone to oral lesions characterized by appearance of soft white slightly elevated plaque most frequently occurring on the buccal mucosa and the tongue and also seen on palate gingival floor of the mouth. The plaque which have been described grossly as resembling milk curds consist chiefly of tangled masses of fungal hyphae with intermingled desquamated epithelium, keratin, fibrin, necrotic debris, leukocytes and bacteria the white plaque can be wipped away with the gauze leaving either a relatively normal appearing mucosa or an erythematous area severe cases entire oral cavity can be involved.

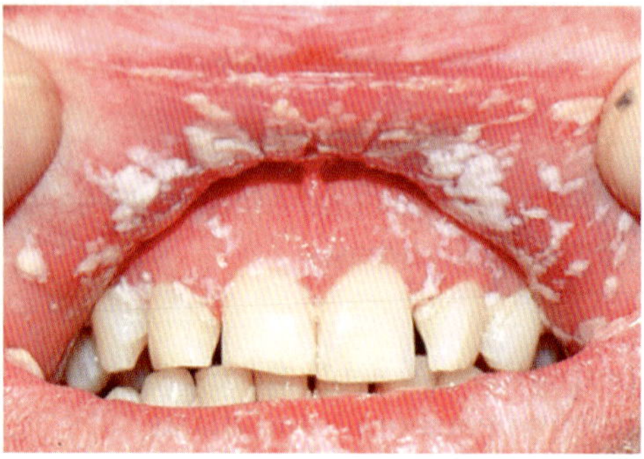

Acute pseudomembranous candidiasis

Acute Atrophic Candidiasis

It is also known as antibiotic sore mouth. It comes under the category of erythematous candidiasis which also include central papillary atrophy of the tongue and kelocandidiasis. It usually occurs as a sequelative course of broad spectrum antibiotics. The lesion in this form of disease appear red or erythematous rather than white thus resembling the pseudomembranous type in which the white membrane has been whipped of it is the only variety of candidiasis which is consistently painful. Central papillary atrophy of the tongue is an asymptomatic symmetric erythematous lesion of dorsal aspect of the posterior region. The erythematous appearance occurs due to loss of fillform papilla.

Chronic Hyperplastic Candidiasis

This is often spoken of as a leukoplakia type of candidiasis. The oral lesion consists of firm, white, persistent plaques, usually on the lips, tongue and cheek. These lesions may persist for period of years. They found a definite relationship between presence of organisms and occurrence of cytologic epithelial atypia biopsy lesions of leukoplakia.

Chronic Mucocutaneous Candidiasis

This is a group of different forms of infection some of which may have multiple features in common although they can be usually separated as entities in general, chronic mucocutaneous candidiasis is characterized by chronic candidal involvement of skin, scalp, nails and mucous membranes. Chronic familial mucocutaneous candidiasis is an inherited disorder probably an autosomal recessive characteristic which occurs early in life usually before the age of 5 and has an equal gender distribution oral lesion occur in these children.

Chronic Localized Mucocutaneous Candidiasis

This is severe of disease also occurring early in life there is no genetic transmission. There is widespread skin involvement in granulomatous and horny masses on the face and scalp. There is increased incidence of other fungal and bacterial infection the mouth is common primary site for the typical white plaques and nail involvement is usually present.

Candidiasis Endocrinopathy Syndrome

This is also a genetically transmitted condition characterized by Candida infection of skin, scalp, nails and mucous membranes classically the oral cavity in association with either hypoadrenalism, hypoparathyroidism, hypothyroidism, ovarian insufficiency or diabetes mellitus. It is recognized that the endocrine manifestations which may be multiple may not appear clinically for several years after the appearance of thrush in the children. Oral findings in the autoimmune polyendocrinopathy-candidiasis syndrome including the common finding of enamel hypoplasia.

Chronic Diffuse Mucocutaneous Candidiasis

Least common form of disease and appears to be of late onset. They exhibit extensive raised crusty sheets involving the limbs, groin, face, scalp and shoulders as well as mouth and nails.

Chronic Atrophic Candidiasis

This is considerd to be synonymous with the condition better known as denture sore mouth. A diffuse inflammation of denture bearing area often occurring with angular cheilitis.

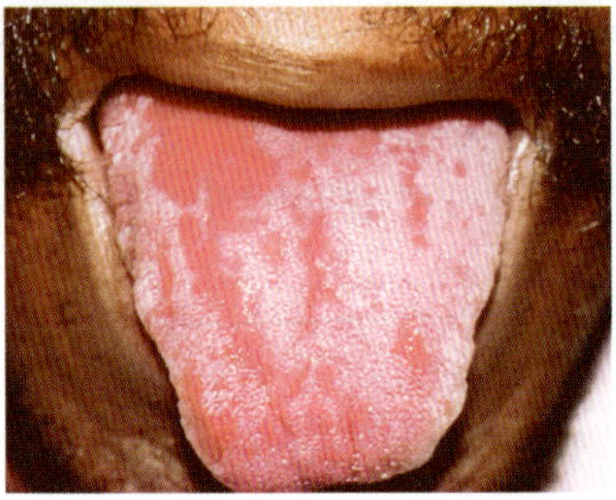

Chronic atrophic candidiasis

ID Reaction

It is the hypersensitivity reaction to candidal antigen which manifest as vesicle and popular rash on the skin of the patients with chronic candidiasis.

Histologic Features

Histologic section of biopsy from a lesion of oral candidiasis will show the presence of yeast cells and hyphae or mycelia in the superficial and deeper layers of involved epithelium.

Treatment

The development of new specific anti fungal agents such as nystatin has been beneficial in the treatment of candidiasis. Other drugs of value are clotrimazoles, amphotericin B and miconazole.

Blastomycosis

North American blastomycosis is a mycotic infection caused by *Blastomyces dermatitidis* and many occur either in a cutaneous form or in a systemic form of involving bones, liver, lungs, subcutaneous tissues, and other organs. The source of the infection in human being is unknown although affected persons commonly work or spend a great deal of time outdoors.

Clinical Features

North American blastomycosis is far more common in men than in women, and typically occurs in middle age. Skin lesions usually begin as small red papules which gradually increase in size and form tiny miliary abscesses or pustules which may ulcerate to discharge the pus through a tiny sinus. The infection commonly spreads through the subcutaneous tissues and become disseminated through the bloodstream.

Oral Manifestations

Oral lesions of blastomycosis may result from either extra pulmonary dissemination or local inoculation with the organism. These lesions may have an irregular erythematous or white intact surface or with ulcerations with irregular rolled borders and varying degrees of pain with a resemblance to squamous cell carcinoma.

Histopathology

This typically shows a mixture of acute inflammation and granulamatous inflammation surrounding variable number of yeasts. These organisms are 8 to 20 microns in diameter.

Treatment

Sulfonamide is used now to treat mild to moderate cases but amphotericin B is used for severe cases. Ketoconazole can also be used.

Paracoccidioidomycosis (South American)

This is a deep fungal infection caused by *Paracoccidioides brasiliensis*. Seen mainly in patients of south America. This has a predilection for males with a 15:1 ratio typically. This is due to protective effect by female hormones.

Clinical Features

Patients are typically middle-aged at the time of diagnosis and most are employed in agriculture. Most cases appear initially as pulmonary infections after exposure to the spores of the organisms. Though infections are self-limiting they may spread by hematogenous or lymphatic spread. A variety of tissues including lymph nodes, skin and adrenal glands.

Oral Manifestations

Mulberry like ulcerations commonly affect alveolar mucosa, gingiva and palate. Lips, tongue, oropharynx and buccal mucosa are also involved in few cases.

Histopathology

Oral lesion may reveal pseudoepitheliomatous hyperplasia and ulceration of oral epithelium. This elicits a granulomatous inflammatory response that is characterized by collection of epithelioid macrophages and multinucleated giant cells scattered large yeast are readily identified after staining with PAS method.

Diagnosis

Demonstration of characteristic multiple budding yeast is adequate to establish diagnosis.

Treatment and Prognosis

Sulfonamides are used in mild to moderate cases. IV Amphotericin B is used in severe cases.

Mucormycosis

This is an opportunistic frequently fulminant fungal infection caused by class *Mucorales*. These are found in a variety of decaying organic materials. Numerous spores are liberated into the air and inhaled by human host. The rhinocerebral form is most relevant to oral health provider.

Mucormycosis is noted in type 1 diabetics who have uncontrolled diabetes and are ketoacidotic. Growth of fungi is enhanced by iron and patients who are taking deferoxamine.

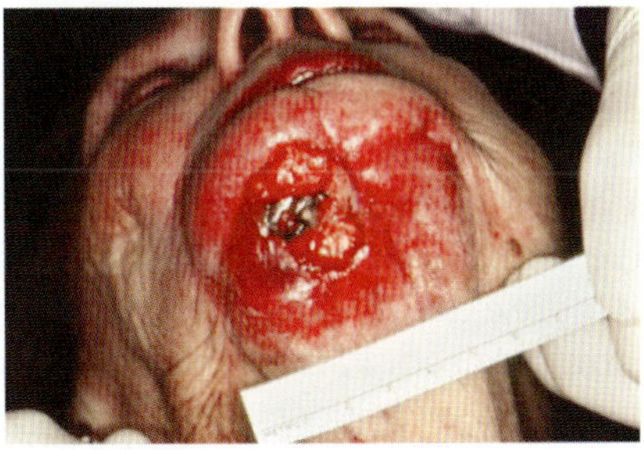

Mucormycosis

Clinical Features

Patients have nasal obstruction, bloody nasal discharge, facial pain, cellulitis and visual disturbances with concurrent proptosis. With progression of disease cranial vault, lethargy, seizures, death occurs.

If maxillary sinus is involved intraoral swelling of maxillary alveolar process, palate or both. If untreated palatal ulceration may evolve appearing black and necrotic. Massive destruction occurs if not treated.

Radiologic Features

Opacification of sinuses may be observed in conjugation with patchy effacement of bony walls of sinuses.

Histopathology

Lesional tissue shows extensive necrosis and numerous large branching non septate hyphae at the periphery. Hyphae tend to branch at 90 degrees. Extensive tissue destruction and necrosis are attributable to preference of fungi for invasion of small blood vessels which disrupts normal blood flow resulting in infarction and necrosis. Neutrophilic infiltrate predominates but host inflammatory response to infection may be minimal.

Treatment

Radical surgical debridement of infection, necrotic tissue and systemic administration of amphotericin B. MRI imaging of head may be useful in determining extent of the lesion.

Aspergillosis

This is a fungal disease characterized by non-invasive and invasive forms. Non-invasive aspergillosis usually affects the normal host appearing as an allergic reaction or a cluster of fungal hyphae. Localised invasive infection of damaged tissue may be seen in a normal host but a more extensive invasive infection is often evident in the immunocompromised. Patients with uncontrolled diabetes are also susceptible to aspergillosis. A species of *Aspergillus* are saprophytic, seen in soil, water or organic debris. Most of the species cannot grow at 37 degrees. Only pathogenic species can replicate at body temperature. The most commonly encountered species in medical setting are *A. flavus* and *A. fumigatus*.

Clinical Features

In normal host the disease may appear as an allergy affecting either sinuses or bronchopulmonary tract. An asthma attack may be triggered by inhalation of spores by a susceptible person. Sometimes a low grade infection occurs in maxillary sinus in the form of fungal hyphae called an aspergilloma. Sometimes this will undergo dystrophic calcification producing a radiopaque body called an antrolith.

Oral Manifestations

Aspergillosis may be encountered after tooth extraction or endodontic treatment especially in maxillary posterior segments. Tissue damage predisposes sinus resulting in localised pain and

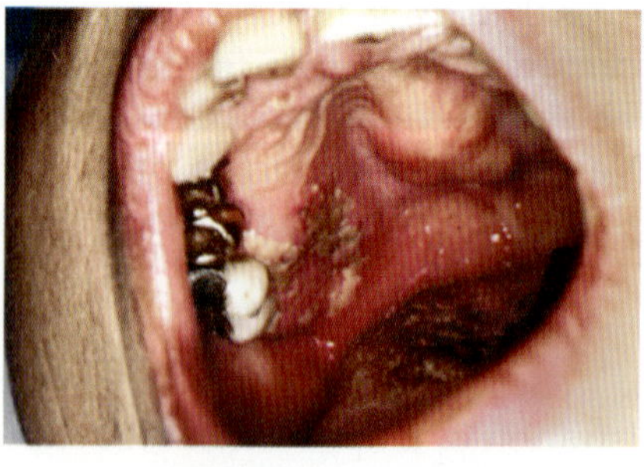

Aspergillosis in palate

tenderness accompanied by nasal discharge. Immunocompromised patients are susceptible to aspergillosis and entry may be marginal gingiva and gingival sulcus. If not treated extensive necrosis with yellow or black ulcers with facial swellings evolve.

Histopathology

Hyphae are branched 3–4 micrometers in diameter. They have a tendency to branch at an acute angle and to involve adjacent blood vessels. Occlusion of vessels results in necrosis in immunocompetent patients a granulomatous inflammatory response in addition to necrosis can be expected.

In non-invasive forms characterized by tangled masses of hyphae with no evidence of tissue infection allergic fungal sinusitis exhibits large pools of eosinophilic inspissated mucin with interspersed sheet like collection of lymphocytes and eosinophils.

Treatment

For immunocompetent patients with non-invasive aspergilloma surgical debridement is done. Patients who have allergic fungal sinusitis are treated with debridement and corticosteroid therapy. For localized invasive forms in immunocompetent host debridement followed by anti fungal medication is done. Prognosis for immunocompetent is much better.

Coccidioidomycosis

Coccidioides immitis is the fungal organism responsible for coccidioidomycosis *C. immitis* is a dimorphic organism, appearing as a mold in its natural environment of the soil and as a yeast in tissues of the infected host.

Clinical Features

- Flulike illness
- Pulmonary symptoms within 1–3 weeks after inhaling the arthrospores
- Fatigue
- Cough

- Chest pain
- Myalgias
- Headache

Chronic Progressive Pulmonary Coccidioidomycosis

It mimics the tuberculosis, clinical presentation are persistent cough, hemoptysis, chest pain, low grade fever and weight loss.

Disseminated Coccidioidomycosis

This occurs when organism spreads hematogenously to extrapulmonary sites.

Clinical Features

Most commonly involved areas include skin, lymph nodes, bone and joints and the meninges.

Treatment

Fluconazole or itraconazole is the drug of choice usually given as high doses for an external period of time.

Ketoconazole also used as drug of choice.

8

Parasitology

DEFINITION

It is the study of parasites which cause disease in man. For centuries parasitic diseases continue to cause high morbidity especially in developing countries. Infection may be low among persons of temperate regions with good hygiene than unhygienic population of tropics. AIDS has given opportunity to parasitic infections which were as a matter of fact rare and less pathogenic.

Poor oral hygiene, warm climate, poor sanitation and non-availability of safe water encourage the parasitic infections.

Parasite

A living organism which get nourishment from another organism where it lives is called parasite which may be:

1. *Ectoparasite:* Those which lives on surface of other organisms, e.g. lice, ticks, mites. It is used to cause an infestation.
2. *Endoparasite:* Those which lives within the body of an organism, e.g. *Entamoeba histolytica*. It is said to cause an infection.
3. *Obligate parasite:* Those who spent some part of their lifecycle in or on host, e.g. *Plasmodium*.
4. *Facultative parasite:* These may be free living but can obtain the nutrition from host too.
5. *Aberrant parasite:* Those parasites which come to a site during migration where it cannot live or develop further.
6. *Accidental parasite:* When parasite attacks unnatural host.

Host

An organism which harbours the parasite is called host.

1. *Definitive host:* Those host which harbours the parasite in adult form or where parasite uses sexual method of reproduction. For example, mosquito is a definite host of malarial parasite.
2. *Intermediate host:* Those which harbours larval stage of the parasite, e.g. man is the intermediate host of malarial parasite.
3. *Paratenic host:* Those in which parasite is viable without development or multiplication. These host help in pass of infection to another host and so-called transport host.
4. *Reservoir host:* Those which makes parasites available parasites for transmission to other host.
5. *Natural host:* Those which are naturally infected with certain species of parasite.

Host Parasite Relationship

1. *Symbiosis:* It is an association that both are dependant on each other that one cannot live without the help of other and none of the partner suffer from the association.
2. *Commensalism:* It is an association in which parasite is deriving benefit without causing injury to its host. It is one form of symbiosis.
3. *Mutualism:* It is an association where parasite and host both are benefited.
4. *Parasitism:* It is an association where parasite gets benefit and host gets nothing in return but always suffers some injury.

Mechanisms of Disease Production by Parasites

A. The Effect of Parasites on their Host

The damage which pathogenic parasite produce on host may be effected in following ways:

1. *Mechanical injury:*
   - It may be inflicted by a parasite by means of pressure as it grows larger, e.g. hydatid cyst.
   - By blockage of lymphatics to produce edema and elephantiasis, e.g. filariasis

- By blockage of blood vessels producing infarction, e.g. *Strongyloides*
- By affecting intestinal tract to produce obstruction, perforation and necrosis, e.g. *Ascaris lumbricoides*
- By destruction of architecture of red blood cells, e.g. *Plasmodium falciparum.*

2. *Deletorious action of toxic substances*
   - In *Plasmodium falciparum* production of toxic substances may cause rigors and other symptoms characteristic of malaria.
   - In *Entamoeba histolytica* lesions in human gut may be due to production of histolytica enzyme.
3. *Deprivation of nutrients and fluids*
   - Parasites may produce disease by competing with the host for food, e.g. *Diphyllobothrium latum* is known to deprive its most of vit B_{12} thus causing megaloblastic anaemia.
   - Certain nematodes and some other blood sucking anthropod parasite quickly lead to severe iron deficiency in patients infested by this, e.g. *Ancylostoma duodenale*
4. *Introduction of pathogenic microorganisms*: Bacterial infection of lesions produced initially by a parasite may occur and in many cases prove lethal, e.g. tetanus is a well-known complication of *Dracunculus medinensis* infection.

B. Reaction of Host to Parasite

The reaction of host to parasite, its developmental stages or products of metabolism may produce two types of host reaction. It may be:

1. Tissue reaction
2. Immunological reaction

1. *Tissue reaction:* Excessive proliferation of certain tissues following invasion by some parasites is known to occur in man, e.g. cirrhosis of liver following deposition of ova of *Schistosoma mansoni*.
2. *Immunological reaction:* Tissue damage may be caused by immunological response to host, e.g. nephritic syndrome by *Plasmodium malariae*.

Classification

1. *Protozoa:* Microscopic, single cell eukaryotes.

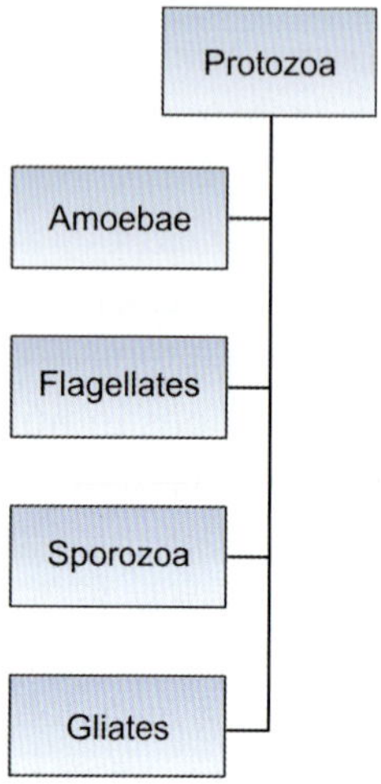

2. *Helminthes:* Macroscopic, multicellular worms with well-differentiated tissues and complex organs.

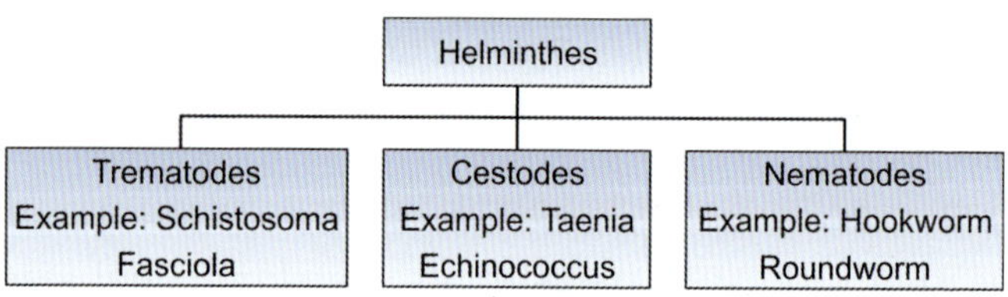

Oral Infections caused by Parasites

Parasite category	*Parasite species*
Arthropods	
Myiasis	*Wohlfahrtia magnifica*
Cestodes	
Teniasis and cysticercosis	*Taenia solium*
	Taenia saginata
Echinococcus	*Echinococcus granulosus*

Nematodes	
Trichinosis	*Trichinella spiralis*
Gongylonema	*Gongylonema pulchrum*
Ascaris	*Ascaris lumbricoides*
Protozoa	
Leishmaniasis	*Leishmania donovani*
Amoebae	*Entamoeba histolytica*
Trichomonas	*Trichomonas tenax*

Arthropods

The phylum arthropods include large number of species than any other phylum. Arthropod= jointed foot

- The phylum arthropoda is large and includes more species than any other phylum.
- Throughout their lifecycle, arthropods
- May host parasitic organisms
- Serve as vector of a wide variety of pathogens
- Cause direct tissue damage and disease in humans.
- *Myiasis*: It is a general term for infection of any organ of the body by maggots that feed on dead or living tissue.
- Acc. To organ infected, myiasis can be
  - Nasopharyngeal
  - Gastrointestinal
  - Auricular
  - Cutaneous.

Maggots may be:

- Parasitic
- Nonparasitic

Myiasis occurs mainly in tropics and is associated with inadequate public and personal hygiene.

Flies that cause myiasis usually have very high fecundity. They lay many eggs from which larvae develop in 3–5 days.

- Species may be–ovoviviparous
  Viviparous
- Myiasis can be–primary myiasis
  Secondary myiasis

Primary myiasis—occurs when larvae penetrate the intact skin or muscle

Secondary myiasis—occurs when open wounds, infections, or burns parasitized

- Normally species of flies favour decaying flesh for larval development but occasionally they deposit eggs or larvae on wounds or ulcers.

Oral Myiasis

Also known as – Dental myiasis
– Gingival myiasis

It is a larval infection of tissues of the oral cavity

It is RARE, and most cases have been reported from developing country.

Predisposing Factors

1. Presence of oral wounds (extraction site, gingivitis, periodontitis)
2. Halitosis
3. History of mouth breathing, especially while sleeping.

It is presumed that the patients become infected with the fly larvae while sleeping.

Occurs in wide age range 9–78 yr

Gender—No gender predilection.

Symptoms

- Pain
- Gingival swelling and bleeding
- "Itchy" feeling in involved side.

Intraoral Examination

- Presence of active larvae burrows into and beneath the gingival leaving an orifice to the burrow, through which the larvae are an exit.

Treatment

- Removal of larvae (ether solution)

- Copious irrigation
- Antibiotics
- Follow up.

Cestodes

Its means girdles or ribbon

- It includes the parasitic tapeworms. Tapeworms have feat, segmented bodies consisting of a head and series of segments known as proglottides.
- The adults of this class are all parasites and exist in the intestinal lumen of vertebrate hosts, including humans.
- Infections with the larval forms, also occurs in humans and cause the diseases known as cysticercosis and echinococcosus.

Teniasis and Cysticercosis

The tapeworm *Taenia saginata* (beef) and *Taenia solium* (pork) produce two different types of disease in humans

- Teniasis
- Cysticercosis

Teniasis: It is an infestation of the human small intestine by adult forms of *T. saginata* or *T. solium*, following ingestion of cysticercus cellulosae cystericercus bovis cysts present in infected or insufficiently cooked pork or beef.

Oral Manifestations

- Red, edematous, hyperplastic mucositis
- Gingival bleeding and pain.

Pathogenesis

Cysticercosis. It results from the extra intestinal encystation of the larval forms of *T. solium* (and rarely *T. saginata*)

- It occurs rarely.

Predisposing Factors

Is a result of poor hygienic conditions, humans may ingest the eggs or proglottides of the parasite.

In the Stomach

1. Gastric secretions free–the embryo (oncospheres)
2. Gain access to the circulation
3. Distribute to various organs and tissues particularly muscle.

In the Tissue

1. Oncospheres develop into the larval forms (cysticercus cellulosae for *T. solium*) (cysticercus bovis for *T. saginata*)
2. Produce cysts
3. Cysticercosis infection.

Common Sites

- Central nervous system
- Ocular structure
- Skeletal muscle
- It is commonly seen in Africa, Asia, India, Latin America, Eastern Europe, and South Africa

Oral Site

Tongue, lips, buccal mucosa.

Gender—No gender predilection

Age—Mean age–22 yr (3–70 yr)

Features—Discrete, firm, non-tender, nodules beneath intact mucosa.

Microscopic Diagnosis

Diagnosis is confirmed by examination of excised cysts showing

- Thin, dense, fibrous connective tissue capsule surrounds a cystic cavity containing a larval form
- Parasites show suckers (hooks) in the cephalic region and digitating walls in a ductal region.

Echinococcosis (Hydatid disease)

Echinococcus granulosus, a parasite of dogs, is one of the smallest tapeworms measuring only 5 mm in length.

In Echinococcus granulosus
Definitive host—Dog and other canine animals
Intermediate host—Man, pig, cattle, horse, goat.

Only the larval form of the echinococcosis infection is found in the human host, and its results in a space occupying lesion known as hydatid cyst.

Hydatid Cyst

Hydatid cyst grows very slowly consist of:

1. Ecto Cyst

It is 1 mm thick. It is a laminated hyaline membrane and has the appearance of a white of a hard boiled egg. It is elastic in nature.

2. Endocyst

It is inner or germinal layer. It is cellular and consists of nuclei dispersed in a protoplasmic mass. It is about 1/4th of mm.

Role

- In formation of outer layer
- Secretion of hydatid fluid
- Formation of blood capsule with scolices.

Hydatid Fluid

- It is secreted by endocyst
- It is clean, colourless, or pale yellow fluid
- It has low specific gravity (1.005–1.010)
- It is slightly acidic (PH "6.7)
- It contains sodium chloride, sodium sulphate, sodium phosphate, and sodium and calcium salts of succinic acid
- Due to its antigenic nature, it is used for casoni's test
- It is highly tonic, when absorbed gives rise to anaphylatic symptoms
- Hydatid cyst fluid contains a granular deposit which settles at the bottom, named a hydatid sand. It consists of blood capsules, free scolices and loose hooklets.

Pathogenesis

The infection is considered serious and has caused fatalities modes of infection:

- By direct contact with infected dogs
- By allowing the dog to feed from the same dish
- By uncooked vegetables contaminated with infected animals feces
- Infection is generally acquired in childhood due to their intimate association with dogs. The cyst may be:
  - Unilocular (e.g. granulosus)
  - Multilocular (e.g. multilocularis).

Oral Manifestations

Sites: Hydatid cysts of the oral structures have reported in tongue, mandible, buccal mucosa, infratemporal fossa.

Clinically: Well-circumscribed, soft or firm, elastic, fluctuant painless swelling. Cysts on tongue cause difficulty with swallowing, speech and mastication as they increased in size. All hydatid cysts should be removed surgically and diagnosed microscopically.

Nematodes

- Nematodes are non-segmented roundworms.
- The members of species mostly live in soil, fresh and salt water.
- Human infections caused by nematodes are among the most prevalent in humans, infecting more than a billion individuals.
- Children are particularly susceptible to these parasites and suffer greater morbidity.
- Nematode infection of the oral cavity is very rare.
- The 10 reports of nematode infection in the dental literature have involved the genera
  - Trichinella
  - Gongylonema
  - Ascaris.

Trichinosis (trichinellosis)

Trichinosis is caused by *Trichinella spiralis* a roundworm parasite capable of infecting all mammals.

Pathogenesis

Modes of infection:

- By poorly cooked meat, usually pork, containing encysted larvae.
- Ingestion of encysted larvae.
- Larvae dissolves in small intestine.
- Adult worm develop, reproduce, and the new larvae enter vessels, circulate and incade striated muscle.
- Muscle cells support the growth and development of the larvae

The "cyst" that develops is actually a misnomer, the cyst is actually a modified, viable cell called the nurse cell. Larvae in the nurse cell can survive for years but eventually but eventually they become calcified and die.

Oral Trichinosis

- Very few cases of oral trichinosis have been reported
- The presence of *T. spiralis* in the oral tissues may have been the result of hematogenous spread of the larvae.
- Zegarelli, et al found encapsulated remnants of *T. spiralis* in biopsies of bilateral soft tissue tumours removed from the mucobuccal fold adjacent to the edentulous maxilla of a 40-year woman.
- Curphey have taken biopsy from a 48-year-old man experiencing recurrent swellings and edentulous right posterior ***alveolar ridge.*** The biopsy revealed hyaline coiled bands, which were diagnosed as *T. spiralis*.
- Miloro and Kinney demonstrated *T. spiralis* in the striated muscle of surgical specimen obtained from bilateral total temporomandibular joint replacements for osteoarthritis and severe segmentative joint disease in a 44-year-old man.
- Bruce and Cheung, et al reported *T. spiralis* in the biopsy specimen of patients undergoing treatment for squamous cell carcinoma raise the issue of carcinogens. The author discussed the possibility of carcinogens and indicated there may be a

relationship between trichinosis and oral squamous. Cell carcinoma.

Gongylonema

Gongylonema is caused by *Gongylonema pulchrum*. It is a small roundworm generally found in the digestive track of domesticated found in farm animals.

There have been 3 reports of this organism in the oral tissues of human.

Two cases were caused by encysted form affecting lower lip and palate.

One case was caused by adult form of the worm affecting buccal vestibule.

Ascaris

Ascaris is caused by *Ascaris lumbricoides*.

- *A. lumbricoides* lives in the upper part of the small intestine, is often referred to as the giant intestine worm because it can grow to a length of more than 30 cm.
- There has been single case report of this worm in the oral structures.
- Hiatt removed the remains of an adult female *A. lumbricoides* from an asymptomatic swelling in the submental area of a 5-year-old boy.
- A persistent thyroglossal duct was considered to the path of entry for the parasite.

Protozoa

Protozoa are a diverse group of eukaryotic, unicellular animals. Parasitic species infect all vertebrate and many invertebrates. The organisms are motile and can be grouped according to their mechanism of locomotion such as cilia, flagella, and pseudopods.

Leishmaniasis

It is a disease caused by Leishmania. These diseases are prevalent where sandfly vectors and mammalian reservoir exist in sufficient numbers to permit frequent transmission. The disease is endemic in the mediterranean basin. Leishmaniasis can be:

- Visceral
- Cutaneous
- Mucosal (rare)

Visceral leishmaniasis (kala-azar)

- It is caused by *Leishmania donovani* in the Indian subcontinent and Africa.
- By *L. infantum* in mediterranean region
- By *L. chagasi* in Americans
- The organism multiplies within the mononuclear phagocytes of the host.

Cutaneous Leishmaniasis

- It is caused by leishmaniasis Mexicans
- The organism multiplies within the phagocytes of skin.

Mucosal Leishmaniasis

- It begins in the nasal mucosa and then involve the oropharynx and larynx.
- Leishmaniasis have considered tropical afflictions and recently were added to the World Health Organization/tropical disease research list of most important diseases.
- They are now also considered a complication of AIDS.

Oral Manifestations

Leishmaniasis may occur as an oral manifestation of either the visceral or the mucosal form of the disease.

1. Abbas, et al reported on 45-year-old man in Sudan with,
   - Swollen easily bleeding gingival
   - The lesions on cheek, uvula and palate were erythematous, red and fissured.
   - Smears of the oral lesions revealed amatigotes and the patients was diagnosed with visceral and mucosal leishmaniasis.
2. Michiels, et al reported oral leishmaiasis and Kaposi's sarcoma in an AIDS patient.

- A biopsy of the tonsil produced a diagnosis of Kaposi's sarcoma in which the leishman–donovan bodies consistent with *L. donovani* were demonstrated.
- Leishmaniasis is an oral disease should be considered as differential diagnosis of oral infection in areas endemic for the disease.

Leishmania braziliensis

- Morphologically it resembles *L. donovani* and has same lifecycle but the vector is forest sandfly.
- Reservoir–small forest rodent.

Clinical Features

- A specific ulcerative granuloma of the skin occurs after incubation period.
- Later the lesion involves mucosa of mouth, nose, pharynx and larynx.
- The lesions may appear as papules, nodules or ulcers.

Amoebae *(Entamoeba gingivalis)*

Morphology

- It is a small amoeba measuring 10–20 micrometer in diameter.
- It is actively motile by multiple pseudopodia.
- It has only the trophozoite stage and there is no cystic phase
- Cytoplasm is divided into clear ectoplasm and granular endoplasm
- Cytoplasmic inclusions consists of bacteria, leucocytes and other substances but never RBCs.
- Nucleus is spherical and has central karyosome
- Nuclear membrane is lined with closely packed chromatin granules.

Sites

- *E. gingivalis* is a parasite of human mouth
- It occurs as a commensal in gingival tissues around the mouth, particularly in the unhealthy tissues such as in pyorrhoea alveolaris.

- It may also occur in apparently healthy mouths and dental plates if they are not cleaned properly
- Also found in the crypts and histological sections and diseased tonsils.

Transmission

By close contact like kissing or from contaminated drinking utensils.

Diagnosis

By demonstration of trophozoites of *E. gingivalis* in the material removed from gingival margins of gums or from denture.

Trichomonas

- The *Trichomonas* are flagellate protozoa with 3.5 anterior flagella.
- The genus *Trichomonas* has been classified into three species according to their habitats.

Trichomonas vaginalis (13 micrometer)

Inhabiting the female genital tract, also found in the urinary tract of females and males.

Trichomonas tenax

- Inhabiting the oral cavity.
- Harmless commensal of oral cavity, periodontal area, carious cavities of tooth, tonsillar crypts, etc.
- *Size:* 5–10 microns
- Transmission through formites, salivary droplets, kissing.

Trichomonas nomini

Inhabiting the ileocaecal region.

Index